Geography
GLOBAL INTERACTIONS
Study and Revision Guide

HL CORE

Simon Oakes

The Publishers would like to thank the following for permission to reproduce copyright material.

Photo credits

p.23 © Motoring Picture Library / Alamy; **p.24** both © Fender image bank; **p.45** © Jemal Countess / Stringer/Getty Images; **p.47** © Kristoffer Tripplaar/Alamy Stock Photo; **p.50** © Luis Sinco / Los Angeles Times / Getty Images; **p.52** © Palash khan/Alamy Stock Photo; **p.57** © Richard Levine / Alamy Stock Photo; **p.61** © Flickr: @John Atherton; **p.63** © Michael Runkel/imageBROKER / Alamy Stock Photo; **p.65** IS489 / Image Source / Alamy Stock Photo; **p.71** © Jeffrey Blackler / Alamy Stock Photo; **p.78** © Bram Janssens/123 RF; **p.83** © TopFoto.co.uk/TopFoto; **p.84** © Anders Peter Photography/Shutterstock.com; **p.90** t © US Air Force Photo/Alamy Stock Photo; **p.90** b © Robert MacPherson/AFP/Getty Images; **p.102** © Tom Wang/Shutterstock.com; **p.107** © David Grossman/Alamy Stock Photo; **p.108** © REMKO DE WAAL/AFP/Getty Images; **p.114** © Xinhua/Alamy Stock Photo

Text credits

p.13 Figure 4.14, Originally published in: Bernice Lee, Felix Preston, Jaakko Kooroshy, Rob Bailey and Glada Lahn 2015, Resources Future; **p.66** Figure 5.25, Mapping Migration © The Economist Newspaper Limited, London (Nov 17th 2011); **p.83** Figure 6.2, Adapted from Hacking Attacks Worldwide; **p.88** Figure 6.8 Adapted from Global giants' tax scheme sandwich leaves bitter taste - The Australian. URL: http://www.theaustralian.com.au/business/global-giants-tax-scheme-sandwich-leaves-bitter-taste/story-e6frg8zx-1226522335217; **p.91** Figure 6.12, source: Branko Milanovic; **p.97** Figure 6.17, source: Nicola Davison, 2016, 'The 22-year-old trying to clean up the Great Pacific Garbage Patch' [Financial Times / FT.com] 04th Aug 2016. Used under licence from the Financial Times. All Rights Reserved.; **p.111** Figure 6.31, source: Peggy Hollinger, 2016. 'Airbus and Boeing put pressure on supply chain' FT.com 26th July 2016. Used under licence from the Financial Times. All Rights Reserved.

Every effort has been made to trace all copyright holders, but if any have been inadvertently overlooked, the Publishers will be pleased to make the necessary arrangements at the first opportunity.

Although every effort has been made to ensure that website addresses are correct at time of going to press, Hodder Education cannot be held responsible for the content of any website mentioned in this book. It is sometimes possible to find a relocated web page by typing in the address of the home page for a website in the URL window of your browser.

Hachette UK's policy is to use papers that are natural, renewable and recyclable products and made from wood grown in sustainable forests. The logging and manufacturing processes are expected to conform to the environmental regulations of the country of origin.

Orders: please contact Bookpoint Ltd, 130 Park Drive, Milton Park, Abingdon, Oxon OX14 4SE. Telephone: (44) 01235 827720. Fax: (44) 01235 400454. Email education@bookpoint.co.uk Lines are open from 9 a.m. to 5 p.m., Monday to Saturday, with a 24-hour message answering service. You can also order through our website: www.hoddereducation.com

ISBN: 9781510403543

© Simon Oakes 2017

First published in 2017 by

Hodder Education,
An Hachette UK Company
Carmelite House
50 Victoria Embankment
London EC4Y 0DZ

www.hoddereducation.com

Impression number 10 9 8 7 6 5 4 3 2 1

Year   2021 2020 2019 2018 2017

All rights reserved. Apart from any use permitted under UK copyright law, no part of this publication may be reproduced or transmitted in any form or by any means, electronic or mechanical, including photocopying and recording, or held within any information storage and retrieval system, without permission in writing from the publisher or under licence from the Copyright Licensing Agency Limited. Further details of such licences (for reprographic reproduction) may be obtained from the Copyright Licensing Agency Limited, Saffron House, 6–10 Kirby Street, London EC1N 8TS.

Cover photo © Hodder & Stoughton Ltd

Illustrations by Aptara, Inc.

Typeset by Aptara, Inc.

Printed in Spain

A catalogue record for this title is available from the British Library.

Contents

How to use this revision and study guide — iv
Features to help you succeed — iv
Getting to know the exam — v
Assessment objectives — v
The examination paper and questions — vi
Understanding and using the PPPPSS concepts — vii

Unit 4 Power, places and networks — 1
- 4.1 Global interactions and global power — 1
- 4.2 Global networks and flows — 12
- 4.3 Human and physical influences on global interactions — 26

Unit 5 Human development and diversity — 42
- 5.1 Development opportunities — 42
- 5.2 Changing identities and cultures — 54
- 5.3 The power of places to resist or accept change — 69

Unit 6 Global risks and resilience — 82
- 6.1 Geopolitical and economic risks — 82
- 6.2 Environmental risks — 94
- 6.3 Local and global resilience — 105

Glossary — 116

How to use this revision guide

Welcome to the *Geography for the IB Diploma Revision and Study Guide*.

This book will help you plan your revision and work through it in a methodological way. The guide follows the Geography syllabus for Paper 3 (Higher Level) topic by topic, with revision and exam practice questions to help you check your understanding.

■ Features to help you succeed

PPPPSS CONCEPTS

These 'think about' boxes pose questions that help you consider and consolidate your understanding and application of the key specialised concepts used in Geography – **place, process, power, possibility**, as well as the two further organising concepts of **scale** and **spatial interactions**. Have a go at every question you come across.

Keyword definitions

The definitions of essential key terms are provided on the page where they appear. These are words that you can be expected to define in exams. A **Glossary** of other essential terms, highlighted throughout the text, is given at the end of the book.

EXAM FOCUS

In the Exam Focus sections at the end of each chapter, example answers to exam-style questions are given and reviewed. Examiner comments and mind maps are used to help you consolidate your revision and practise your exam skills.

■ CHAPTER SUMMARY KEY POINTS

At the end of each chapter, a knowledge checklist helps you review everything you have learned over the previous pages. An Evaluation, Synthesis and Skills (ESK) summary is also included. This helps show how the knowledge you have acquired may be applied in order to analyse information, evaluate issues and tackle big geographic questions.

You can keep track of your revision by ticking off each topic heading in the book. There is also a checklist at the end of the book. Use this checklist to record progress as you revise. Tick each box when you have:

- revised and understood a topic
- read the exam-style questions in the Exam focus sections, completed the activities and reviewed any example answer comments.

Use this book as the cornerstone of your revision. Don't hesitate to write in it and personalize your notes. Use a highlighter to identify areas that need further work. You may find it helpful to add your own notes as you work through each topic. Good luck!

Getting to know the exam

Exam paper	Duration	Format	Topics	Weighting	Total marks
Paper 1 options (SL)	1 hour 30 mins	Structured questions and essays	2 options	35	40
Paper 1 options (HL)	2 hours 15 mins	Structured questions and essays	3 options	35	60
Paper 2 core (SL/HL)	1 hour 15 mins	Structured questions and essay	All	40 (SL) 25 (HL)	50
Paper 3 core (HL only)	1 hour 45 mins	Extended writing and essay	All	20	28

At the end of your Geography course, you will sit two papers at SL (Paper 1 and Paper 2) and three papers at HL (Paper 1, Paper 2 and Paper 3). These external exams account for 80% of the final marks at HL and 75% at SL. The other assessed part of the course (20% at HL and 25% at SL) is the Internal Assessment which is marked by your teacher, but externally moderated by an examiner. Here is some general advice for the exams:

- Make sure you have learned the command terms (e.g. evaluate, explain, outline, etc.); there is a tendency to focus on the content in the question rather than the command term, but if you do not address what the command term is asking of you, then you will not be awarded marks. Command terms are covered below.
- If you run out of room on the page, use continuation sheets and indicate clearly that you have done this on the cover sheet.
- The fact that the answer continues on another sheet of paper needs to be clearly indicated in the text box provided.
- Plan your answer carefully *before* you begin your extended writing and essay tasks.
- Spend time learning the key terms featured in the Guide as these words may feature as part of the essay titles in your examination.
- Get to know the specialised concepts (place, process, power, possibility) and organising concepts (scale and spatial interactions). Answers that are awarded the highest marks are likely to make use of these concepts (see page vii).

Assessment objectives

To successfully complete the course, you have achieved certain assessment objectives. The following table shows all of the command terms used in Paper 3 which this book supports, along with an indication of the depth required from your written answers.

Analyse	AO2	**Assessment objective 2** Demonstrate application and analysis The part (a) question of your HL extension exam is primarily an AO2 task.	These command terms require students to apply their knowledge and understanding to a well-defined task such as, 'Using examples, explain why glocalization is an important strategy for global businesses.'
Distinguish	AO2		
Explain	AO2		
Suggest	AO2		

Discuss	AO3	**Assessment objective 3** Demonstrate synthesis and evaluation The part (b) question of your HL extension exam is primarily an AO3 task.	These command terms require students to rearrange a series of geographic ideas, concepts or case studies into a new whole and to provide evaluation or judgements based on evidence. For example: 'Discuss the view that globalization brings more costs than benefits to societies.'
Evaluate	AO3		
Examine	AO3		
To what extent	AO3		

The table on the next page defines command words used most commonly in the Geography Paper 3 examination.

Term	Definition
Analyse (AO2)	Break down in order to bring out the essential elements or structure.
Explain (AO2)	Give a detailed account, including reasons or causes.
Examine (AO3)	Consider an argument or concept in a way that uncovers the assumptions and inter-relationships of the issue.
Discuss (AO3)	Offer a considered and balanced review that includes a range of arguments, factors or hypotheses. Opinions or conclusions should be presented clearly and supported by appropriate evidence.
To what extent (AO3)	Consider the merits or otherwise of an argument or concept. Opinions and conclusions should be presented clearly and supported with empirical evidence and sound argument.

It is essential that you are familiar with these terms, so that you are able to recognize the type of response you are expected to provide.

The examination paper and questions

You choose one from three optional questions in the examination. Each optional question has two parts:

- part (a) extended writing task with a maximum mark of 12
- part (b) essay with a maximum mark of 16.

The **part (a) extended writing task** asks you to **explain** or **analyse** a geographic issue, theory or idea. It is an **AO2** task and does not require you to argue a viewpoint or arrive at a conclusion. You should spend **20–25** minutes on this task. Examples of part (a) questions include:

(a) Analyze how powerful states influence global interactions in ways which benefit themselves. (See pages 10–11 for a worked example of this.)

(a) Explain why it might be hard to establish how human development varies between different countries. (See page 53 for a worked example of this.)

A simplified version of the levels-based mark scheme for part (a) questions looks like this:

Mark scheme

1–3 marks	Response is general, not focused on the question, and lacks detail and structure.
4–6 marks	Response only partially addresses the question; evidence is both relevant and irrelevant and is largely unstructured.
7–9 marks	Response addresses most parts of the question, outlines an analysis supported by relevant evidence but may lack clear links between paragraphs.
10–12 marks	Response addresses all aspects of the question; the analysis is explained using evidence integrated in the paragraphs, and it is well structured.

The **part (b) essay** task requires you to **discuss**, **examine** or **evaluate** a statement. It is an **AO3** task, which requires you to write critically and in a conceptually informed way about the question statement. Alternatively, you may be required to argue for and against a viewpoint before arriving at a conclusion. You should spend **35–40 minutes** on this task. Examples of part (b) questions include:

(b) Examine the interactions between technology, transnational corporations and the growth of globalization. (See page 39 for a worked example of this.)

(b) Barriers to globalization are on the rise in many parts of the world. Discuss this statement. (See page 80 for a worked example of this.)

A simplified version of the levels-based mark scheme for part (b) questions looks like this:

Mark scheme

1–4 marks	Response is general, not focused on the question, and lacks detail and structure. No synthesis or evaluation expected.
5–8 marks	Response only partially addresses the question with limited links to the guide; evidence is both relevant and irrelevant and is largely unstructured. No synthesis or evaluation expected.
9–12 marks	Response addresses most parts of the question with developed links to the guide and outlines an analysis supported by relevant evidence. Synthesis OR evaluation required.
13–16 marks	Response addresses all aspects of the question, analysis is explained and evaluated using evidence integrated in the paragraphs. Synthesis AND evaluation required.

Understanding and using the PPPPSS concepts

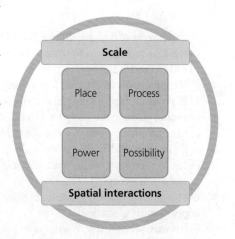

An important feature of the new 2017 Geography course is the inclusion of specialized and organizing Geography Concepts. These are shown in the diagram below.

Place	A portion of geographic space, which is unique in some way. Places can be compared according to their cultural or physical diversity, or disparities in wealth or resources. The characteristics of a place may be real or perceived.
Process	Human or physical mechanisms of change, such as migration or erosion. Processes operate on varying timescales. Linear systems, circular systems and complex systems are all outcomes of the way in which processes operate and interact.
Power	The ability to influence and affect change or equilibrium at different scales. Power is vested in citizens, governments, institutions and other players, and in processes in the natural world. Equity and security, both environmental and economic, can be gained or lost as a result of the interaction of powerful forces.
Possibility	Alternative events, futures and outcomes that geographers can model, project or predict with varying degrees of certainty. Key contemporary possibilities include the degree to which human and environmental systems are sustainable and resilient, and can adapt or change.
Scale	Places can be identified at a variety of geographic scales, from local territories to the national or state level. Global-scale interactions occur at a planetary level.
Spatial interactions	Flows, movements or exchanges that link places together. Interactions may lead places to become interdependent on one another.

Essentially, these six ideas help provide you with a roadmap to 'thinking like a geographer'. For instance, how might a geographer approach answering a very general essay question such as:

'To what extent does a global culture exist?'

This is a very broad question and as a result potentially tricky to answer well. The specialized and organizing Geography concepts can be used to help you 'scaffold' your answer. Familiarity with the PPPPSS framework provides you with the basis for a series of further questions you may want to address as part of your overall answer, as follows:

- Are elements of a global culture found in every **place** or only in some places?
  (This may prompt you to write about poor communities living in isolated islands in Indonesia.)
- What **processes** lead to the formation of a global culture?
  (This may prompt you to write about a process such as 'McDonaldization'.)
- Do some places have the power to resist the spread of a global culture?
  (This may prompt you to write about North Korea, a politically isolated state.)
- Is there a **possibility** that global culture will spread even further?
  (This may prompt you to write about current affairs such as the new wave of nationalism sweeping across Europe.)
- Do some local-**scale** communities resist global culture even in states that appear highly globalized?
  (This may prompt you to write about rural communities in the USA or Europe.)
- What different kinds of **spatial interaction** can contribute to the spread of global culture?
  (This may prompt you to write about how flows of migrants, commodities and ideas can all contribute to the diffusion of culture.)

In order to write a good essay, you do not need to do all of this, of course. But it may be helpful to try to draw on two or three of the specialized and organized concepts when planning an essay. You can also make use of the Group 3 concept of **perspectives**. Most questions can be debated from the varying and sometimes contrasting perspectives of different **stakeholders**, or players (all of whom may be located at local, national or global scales).

Unit 4 Power, places and networks

4.1 Global interactions and global power

Revised

This study of **global interactions** has a broader perspective than some more conventional studies of globalization that often emphasize a linear process involving the domination and imposition of Western economic, political and cultural models and values on the world.

In the context of this course, global interactions suggests a two-way and complex process whereby dominant economic commodities and conventions, cultural traits, social norms and global political frameworks may be adopted, adapted or resisted by local societies. Central to this analysis is an appreciation of power:

- Some powerful people and places bring changes to other individuals and societies at a global scale.
- Local societies and stakeholders differ in their power to resist or adapt to globalizing forces and risks.

> **Keyword definition**
>
> **Global interactions** This phenomenon includes all of the varied economic, social, political, cultural and environmental processes that make up globalization. It also encompasses the many local opposition movements and new cultural forms that result when globalizing forces meet and interact with local societies and stakeholders.

Analysing globalization

Revised

The umbrella term **globalization** is used to describe a variety of ways in which places and people are now more connected with one another than they used to be. Many differing definitions of the term are in use (Figure 4.1).

- The words used reflect the varying perspectives of the writers: some definitions are primarily economic, such as the statement by the International Monetary Fund (IMF), which views globalization as: 'The growing interdependence of countries worldwide through the increasing volume and variety of cross-border transactions in goods and services and of international capital flows, and through the more rapid and widespread diffusion of technology.'
- Other definitions, as you can see, put greater emphasis on the cultural and political transformations that are also part of the globalization process (Figure 4.2).
- Some definitions are critical: this is because some people and organizations believe that recent global-scale changes brought by **transnational corporations TNCs)** and governments are often deeply harmful to people, places and environments.

> **PPPPSS CONCEPTS**
>
> Think about how the concept of place can be applied at varying scales in the country where you live (for example, neighbourhood, town, city, country).

> **Keyword definition**
>
> **Transnational corporations (TNCs)** Businesses whose operations are spread across the world, operating in many nations as both makers and sellers of goods and services. Many of the largest are instantly recognizable 'global brands' that bring cultural change to the places where products are consumed.

A rapid and huge increase in the amount of economic activity taking place across national boundaries has had an enormous impact on the lives of workers and their communities everywhere. The current form of globalization, with the international rules and policies that underpin it, has brought poverty and hardship to millions of workers, particularly those in developing and transition countries. *UK Trade Union Congress*

Globalization is a process enabling financial and investment markets to operate internationally, largely as a result of deregulation and improved communications. *Collins Dictionary*

The term 'globalization' refers to the increasing integration of economies around the world, particularly through the movement of goods, services, and capital across borders. It refers to an extension beyond national borders of the same market forces that have operated for centuries at all levels of human economic activity – village markets, urban industries, or financial centres. There are also broader cultural, political, and environmental dimensions of globalization. *IMF*

Globalization can be conceived as a set of processes which embodies a transformation in the spatial organization of social relations and transactions, expressed in transcontinental or interregional flows and networks of activity, interaction and power. *Held and McGrew (Globalization Theory)*

The expansion of global linkages, organization of social life on a global scale, and growth of global consciousness, hence consolidation of world society. *Frank Lechner (The Globalization Reader)*

It might mean sitting in your living room in Estonia while communicating with a friend in Zimbabwe. It might mean taking a Bollywood dance class in London. Or it might be symbolized in eating Ecuadorian bananas in the European Union. *World Bank (for schools)*

A social process in which the constraints of geography on economic, political, social and cultural arrangements recede, in which people become increasingly aware that they are receding, and in which people act accordingly. *Malcolm Waters (Globalization)*

Figure 4.1 Defining globalization

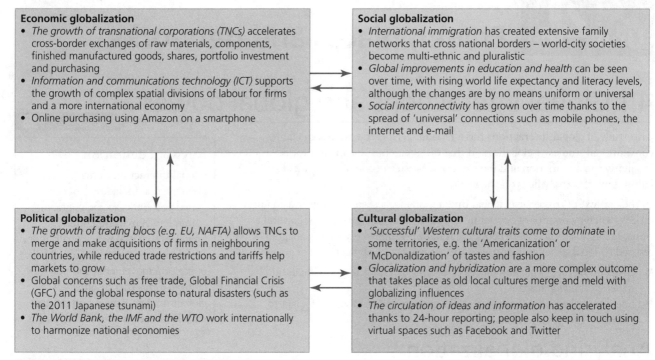

Figure 4.2 Four interconnected strands of globalization

Modern globalization does not lack in history. It is the continuation of a far older, ongoing economic and political project of global trade and empire building. Economies have been interdependent to some extent since the time of Earth's first great civilizations, such as ancient Egypt, Syria and Rome. There is certainly nothing novel in the global power-play and ambition of strong people, nations and businesses; globalization may be regarded as the latest chapter in a long story of globally connected people and places.

In the past, global connections were achieved through:

- **trade** – especially after 1492 when Columbus reached the Americas and the traditional world economy began to take shape
- **colonialism** – by the end of the nineteenth century, the British Empire directly controlled one quarter of the world and its peoples
- **cooperation** – ever since the First World War ended in 1918, international organizations similar to today's United Nations (UN) have existed.

Modern (post-1940s) globalization differs from the global economy which preceded it due to the following factors.

- **Lengthening** of connections between people and places, with products shipped greater distances than in the past, and tourists travelling further from home.
- **Deepening** of connections, with the sense of being connected to other people and places now penetrating more deeply into almost every aspect of life. Think about the food you eat each day and the many places it is sourced from. It is difficult not to be connected to other people and places through the products we consume.
- **Faster speed** of connections, with people able to talk to one another in real-time, using technologies such as Skype, or travel very quickly between continents using jet aircraft. Distance between continents has become measured in mere hours of flight time. Air travel, the telephone, internet access and containerized shipping are among the crucial new technologies that help coordinate economic, political, cultural and sporting activities taking place simultaneously in different parts of the world. In virtual environments such as Facebook, truly globalized communities work and play in perfect synchronicity.

Figure 4.3 shows a timeline of the post-1945 period. It includes key economic, political and technological influences or events that have helped mould the world's states into a single, densely interconnected economic unit.

1944–45 Establishment of the World Bank, International Monetary Fund (IMF) and origin of the World Trade Foundation (WTO). The post-war Bretton Woods conference provides the blueprint for a free-market non-protectionist world economy where aid, loans and other assistance become available for countries prepared to follow a set of global financial guidelines written by powerful nations.

1960s Heavy industry in the 'Western' developed economies of Europe and America is increasingly threatened by rising production in South East Asia, including Japan and the emerging Asian Tiger economies, notably South Korea. Unionized labour costs push up the price of production for Western shipbuilding, electronics and textiles. Most advanced economies enter a period of falling profitability for industry.

1970s The OPEC Oil Crisis of 1973 puts pressure on Western industry. Rising fuel costs trigger a 'crisis of capitalism' for Europe and America, whose firms begin out-sourcing and off-shoring their production to low labour-cost nations. Meanwhile, soaring petrodollar profits for Middle East OPEC nations signal that the United Arab Emirates and Saudi Arabia are on their way to becoming new global hubs. In 1978, China begins economic reforms and opens up its economy.

1980s Financial deregulation in major economies like the United Kingdom and the United States brings a fresh wave of globalization, this time involving financial services, share dealing and portfolio investment (by 2008, financial markets will have an inflated value more than twice the size of actual world GDP!). The collapse of the Soviet Union in 1989 significantly alters the global geopolitical map, leaving the United States as the only 'superpower'.

1990s Landmark decisions by India (1991) and China (1978) to open up their economies bring further change to the global political map. Established powers strengthen their regional trading alliances, including the European Union (EU, 1993) and North American Free Trade Association (NAFTA, 1994). The late 1990s Asian financial crisis is an early warning of the risks brought by loosely regulated free market global capitalism.

2000s Major flaws in the globalized banking sector emerge during the 2007–09 Global Financial Crisis (GFC). Unsecured loans totalling trillions of dollars undermine leading banks. This brings a negative multiplier effect that causes a fall in the value in global gross domestic product (GDP). In an interconnected world, growth slows for the first time in two decades for China and India, the two great out-sourcing nations and now emerging superpowers.

2010s Growth remains slow following the GFC, with many countries slipping in and out of recession, including Russia and Brazil. Problems in Greece and Portugal escalate into the Eurozone crisis. Despite slower growth, China overtakes the United States to become the largest economy by purchasing power parity (PPP). In many countries, popular opposition to migration and free trade is on the rise and the United Kingdom votes to leave the EU. However, global internet and social networking use reaches new record levels. Experts struggle to tell whether globalization is increasing, pausing or retreating.

TIME

Figure 4.3 A post-1945 timeline for economic and political globalization

A sense of global connectivity is not yet shared by everyone in the world though. Some nations and regions (for example, parts of the Sahel) experience a much more 'shallow' form of integration (Figure 4.4). There can also be great disparity among a country's citizens in terms of how 'global' they feel. For instance, many citizens in Brazil's core cities of Rio and São Paulo are globally connected, either as producers of goods or as consumers of the football World Cup held in Brazil in 2014. However, some Amazon rainforest tribes, such as the Korubo people, have little or no knowledge of the outside world, and lack much connectivity with other societies.

> **PPPPSS CONCEPTS**
>
> Think about how all states are globally connected (though to varying degrees) but some smaller-scale places within them may not be.

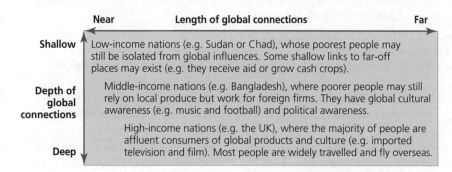

Figure 4.4 Degrees of engagement with global interactions

Unit 4 Power, places and networks

Measuring globalization

Various efforts have been made to quantify the extent to which different countries participate in global interactions. Uneven levels of globalization can be measured using indicators and indices.

The Swiss Institute for Business Cycle Research, also known as KOF, produces an annual Index of Globalization. In 2014, Ireland and Belgium were the world's most globalized countries according to the KOF index (Figure 4.5). A complex methodology informs each report (Table 4.1). Levels of economic globalization are calculated by examining trade, FDI figures and any restrictions on international trade. Political globalization is also factored in, for instance, by counting how many embassies are found in a country and the number of UN peace missions it has participated in. Finally, social globalization is accounted for, defined by KOF as 'the spread of ideas, information, images and people'. Data sources for this include levels of internet use, TV ownership, and imports and exports of books.

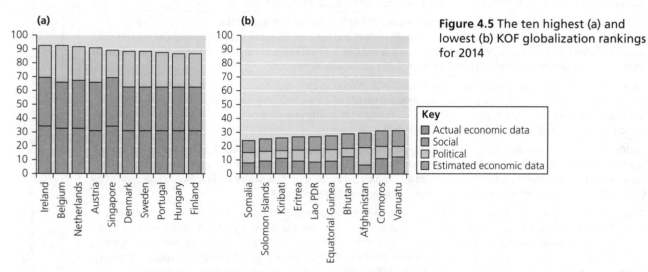

Figure 4.5 The ten highest (a) and lowest (b) KOF globalization rankings for 2014

Table 4.1 Calculating a country's globalization score using the KOF index

Stage	
1	To study economic globalization, information is collected showing long-distance flows of goods, capital and services. Data on trade, FDI and portfolio investment are studied, including figures from the World Bank.
2	A second measure of economic globalization is made by examining restrictions on trade and capital movement. Hidden import barriers, mean tariff rates and taxes on international trade are recorded. As the mean tariff rate increases, countries are assigned lower ratings.
3	Political globalization data are looked at next. To proxy the degree of political globalization, KOF records the number of embassies and high commissions in a country as well as the number of international organizations of which the country is a member and also the number of UN peace missions a country has participated in.
4	The assessment of social globalization starts with use of personal contacts data. KOF measures direct interaction among people living in different countries by recording: a) international telecom traffic (traffic in minutes per person); b) tourist numbers (the size of incoming and outgoing flows); and c) the number of international letters sent and received.
5	The study of social globalization also requires use of information flows data. World Bank statistics are employed to measure the potential flow of ideas and images. National numbers of internet users (per 1,000 people) and the share of households with a television set are thought to show 'people's potential for receiving news from other countries – they thus contribute to the global spread of ideas'.
6	Finally, 'cultural proximity' data are collected. By KOF's own admission, this is the dimension of social globalization most difficult to measure. The preferred data source is imported and exported books. 'Traded books,' explains KOF, 'proxy the extent to which beliefs and values move across national borders.'
7	Now all the data have been collected, a total of 24 variables (covering economic, political and social globalization) are each converted into an index on a scale of 1 to 100, where 100 is the maximum value for each specific variable over the period 1970–2006 and 1 is the minimum value. High values denote greater globalization. However, not all data are available for all countries and all years. Missing values are substituted by the latest data available. Averaging then produces a final score out of 100.
8	Each year's new KOF index scores are added to a historical series covering more than 30 years, beginning in 1970. Changes in globalization over time can then be studied.

While there is obvious merit in KOF's multistrand approach to measuring globalization, there are considerable grounds for criticism too. For instance, does possession of a television set really make a family more globalized? Equally, the reasons why some countries volunteer large numbers of troops for US missions are complex; it need not necessarily be the case that the most economically globalized countries are the most militarily proactive (take Japan and Germany, for instance).

Another well-known measure of globalization is the AT Kearney World Cities Index. New York, London, Paris, Tokyo and Hong Kong are all ranked highly as 'Alpha' global hubs for commerce. The ranking is established by analysing each city's 'business activity', 'cultural experience' and 'political engagement'. The data supporting this include a count of the number of TNC headquarters, museums and foreign embassies, respectively.

The KOF Index and Kearney Index combine many data sources, which can be critiqued on the grounds of either reliability or validity. Some data suffer from crude averaging and statistical gaps. Other indicators are arguably poor proxies for globalization (hours spent watching TV, for instance). While they are an interesting starting point for the study of globalization, both indexes lack the rigour and trustworthiness of, say, properly sampled and peer-reviewed scientific surveys of climatic data for different countries.

Global superpowers

Revised

The term global superpower was used originally to describe the ability of the USA, USSR and the British Empire to project power and influence anywhere on Earth to become a dominant worldwide force.

- The British Empire was a colonial power, alongside France, Spain, Portugal and other European states. Between approximately 1500 and 1900, these leading powers built global empires. One result was the diffusion of European languages, religions, laws, customs, arts and sports on a global scale.
- In contrast to the direct rule of the British in the 1800s, the USA has dominated world affairs since 1945 mainly by using indirect forms of influence or **neo-colonial** strategies. These include the US government's provision of international aid and the cultural influence of American media companies (including Hollywood and Facebook). Alongside such **soft power** strategies, the US government has routinely made use of **hard power**. This means the geopolitical use of military force (or the threat of its use) and the economic influence achieved through forceful trade policies, including economic sanctions or the introduction of import tariffs. The term 'smart power' is used to describe the skilful combined use of both hard and soft power in international relations (Figure 4.6).

Keyword definitions

Neo-colonial The indirect actions by which developed countries exercise a degree of control over the development of their former colonies. This can be achieved through varied means including conditions attached to aid and loans, cultural influence and military or economic support (either overt or covert) for particular political groups or movements within a developing country.

Soft power The political scientist Joseph Nye coined the term 'soft power' to mean the power of persuasion. Some countries are able to make others follow their lead by making their policies attractive and appealing. A country's culture (arts, music, cinema) may be viewed favourably by people in other countries.

Hard power This means getting your own way by using force. Invasions, war and conflict are very blunt instruments. Economic power can be used as a form of hard power: sanctions and trade barriers can cause great harm to other states.

PPPPSS CONCEPTS

Think about what the word power means at different scales and in different contexts. What can make a person powerful, for instance?

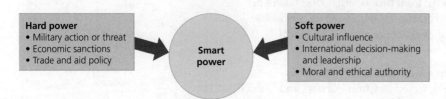

Figure 4.6 The ingredients of 'smart power'

Other than the USA, what other states can claim to be true global superpowers?

- China became the world's largest economy in 2014 and exerts power over the global economic system through its sheer size. Other **emerging economies** including India and Indonesia play an increasingly important global role.
- Russia uses military force and exports of gas to maintain influence.
- Although no single European country can equal the influence of the USA, several have remained significant global players in the post-colonial world (notably the G8 nations of Germany, France, Italy and the UK). Another view is that European states can only rival the USA's global superpower status when they work together as members of the European Union.

> **Keyword definition**
>
> **Emerging economy** Countries that have begun to experience higher rates of economic growth, often due to rapid industrialization. Emerging economies correspond broadly with the World Bank's 'middle-income' group of countries and include China, India, Indonesia, Brazil, Mexico, Nigeria and South Africa.

Figure 4.7 provides an assessment of several influential countries using a range of criteria to create a superpower index.

Constructing a superpower index

Data on superpower characteristics can be used to quantify their power and influence. Single measures, such as population size or total GDP, are unlikely to be good indicators because they only quantify one aspect of power.

Figure 4.7 shows a power index using four quantitative measures:

- total GDP (economic)
- total population (resources/demographic)
- nuclear warheads (military)
- TNCs (economic/cultural).

Each measure has been ranked, with 1 being the highest score. Total GDP and TNCs have been scaled (multiplied by 3 and 2, respectively) to reflect their greater importance as measures of power.

Using this index, the USA has a total score of 10 and is the most powerful country. China's score of 14 puts it close to the USA. Other countries are some distance behind these scores. This reflects the fact that they are powerful in some areas but not others. A wide range of data can be used to construct power indices, although the results are often surprisingly similar.

	Total GDP, US$ billion	Rank, scaled × 3	Total population, million	Rank	Nuclear warheads	Rank	Fortune Global 500 TNCs	Rank, scaled × 2	Sum of ranks
China	10.4	6	1,360	1	260	3	98	4	14
India	2.1	15	1,250	2	120	5	7	10	32
Japan	4.6	9	126	5	0	6	54	6	26
Russia	1.9	18	146	4	4,700	1	5	12	35
UK	3.0	12	64	6	215	4	29	8	30
USA	17.4	3	320	3	4,500	2	127	2	10

Source: Cameron Dunn

Figure 4.7 Constructing a superpower index (2014 data)

Many countries can make a claim to be globally powerful in certain ways. For example, the tiny Middle Eastern state of Qatar has proved to be capable of exerting great influence around the world. It has the highest per capita GDP in the world, in excess of US$100,000. Its wealth and global influence, like neighbouring Saudi Arabia's, are partly due to fossil fuel wealth: Qatar has 14 per cent of all known gas reserves. Qatar's government has reinvested its petrodollar wealth in ways that have diversified the national economy and built global influence too:

- The city of Doha has become a powerful place where international conferences and sporting events are held, served by Qatar Airways and Doha International Airport. Important UN and World Trade Organization (WTO) meetings have taken place in Doha including the 2012 UNFCCC Climate Negotiations. The city is set to host the 2022 football World Cup.
- Qatar's Al Jazeera media network rivals the BBC and CNN for influence in some parts of the world and is an important source of soft power.

However, many people regard Qatar as a regional power, rather than a true global superpower.

CASE STUDY

RIVAL SUPERPOWERS? CHINA AND THE USA
The USA and China are the world's leading superpowers.

- Until recently, the USA was widely seen as being the world's unchallenged 'hyperpower', meaning that it was dominant in all aspects of power.
- Since the GFC, however, it has become harder to sustain this unipolar view of the world. China is increasingly viewed as rivalling the USA in many ways (Figure 4.8).
- Up until very recently, the USA benefited most from globalization by being able to regulate the terms of its own global interactions with other countries and TNCs in ways that returned significant economic and political rewards. For this reason, conventional wisdom used to be that globalization and 'Americanization' were one and the same thing. However, the roadmap is rapidly changing, as Table 4.2 shows.

Figure 4.8 China and the USA are vying for leading superpower status

Table 4.2 Analysing and evaluating the superpower status and global influence of the USA and China

	Analysis	Evaluation
USA	• The 320 million people who live here (less than one-twentieth of the world's population) own more than 40 per cent of global personal wealth. Of the 500 largest global companies, one quarter were US-owned in 2015. • The USA has disproportionate influence over important intergovernmental organizations, many of which are based in Washington DC or New York. To date, all World Bank presidents have been US citizens. • US cultural influence is so strong that terms like 'Americanization' and 'McDonaldization' are widely used to describe the way US food, fashion and media have shaped global culture. • The USA has used a combination of overt military power and covert intelligence operations to intervene in the affairs of almost 50 states since 1945, according to William Blum, a former US State Department official.	• The USA's influence over international organizations, including the UN, NATO, the IMF and the World Bank, has given it greater influence over global politics than any other state. The USA was the main architect of the global economic system created at the end of the Second World War, and the principles that inform it. Over time, the free market philosophy that institutions like the World Bank and International Monetary Fund promote have become known as 'the Washington consensus'. • The multiple ways in which the USA is able to influence global interactions make it a true global superpower (some people have even dubbed the USA a 'hyperpower'). No other country has such a formidable combination of geopolitical, economic and cultural tools at its disposal. • President Trump's view is that the USA has lost too much influence to China and that the USA is no longer powerful enough.
China	• Prior to 1978, China was a poor and politically isolated country, 'switched-off' from the global economy. Under the communist leadership of Chairman Mao Zedong, millions had died from famine. Most people lived in poverty in rural areas. This changed in 1978 when Deng Xiaoping began the radical 'Open Door' reforms that allowed China to embrace globalization while remaining under one-party authoritarian rule. • Today, China is the world's largest economy. Over 400 million of its people are thought to have escaped poverty since the reforms began. FDI from China and its TNCs is predicted to total US$1.25 trillion between 2015 and 2025.	• The average income of China's population is still less than one-third of US citizens. Further catch-up may be slow now economic growth has begun to tail off in China. • The country faces the rising challenge of an ageing population (a legacy of its one-child policy, which was only abandoned recently). • China lacks the soft power of the USA, in part due to its cultural isolation from the rest of the world (few foreign films are allowed into China and internet freedoms are restricted by its government). The lack of democracy in China also adversely affects its international relations with some other countries.

Powerful global organizations and groups

Revised

The world map of power is complicated. It is not only the governments of individual states that influence the world economy and global interactions. Global organizations (including lending institutions) and global groups of nations play an important role too.

The UN was the first post-war intergovernmental organization (IGO) to be established. The UN's contribution to global development is vast. No other international organization has the same degree of influence over global interactions. Over time, its remit has grown to include human rights, the environment, health and economics. The General Assembly is made up of voting representatives for all 193 member states. Important UN achievements have included:

- the Declaration of Human Rights and the International Court of Justice
- the 1992 Conference on Environment and Development (the 'Earth Summit')
- the Millennium Development Goals (2000) and the Sustainable Development Goals (2015).

Alongside the UN, other important organizations and groups have a significant role in global governance, particularly in relation to matters of trade. Table 4.3 analyses and evaluates the work done by global lending institutions, while Table 4.4 shows several important and powerful global groups. Figure 4.9 shows the relationship between some of these groups and the wealth of their members.

> **Keyword definition**
> **BRICS group** Brazil, Russia, India, China and South Africa are five countries whose economies were growing rapidly in the early years of the twenty-first century.

Table 4.3 Global lending institutions

Institution	Analysis	Evaluation
International Monetary Fund (IMF)	• The IMF monitors the economic development of countries. Under the umbrella of the UN, it lends money to states in financial difficulty which have applied for assistance. • Help is provided to countries across the development spectrum when they encounter financial difficulty. Between 2010 and 2015, almost US$40 billion was lent to Greece to help end a period of financial crisis. • The IMF has always had a European president but is based in Washington DC.	• IMF rules and regulations are controversial, especially the strict conditions imposed on borrowing governments. In return for help, recipients agree to run free-market economies that are open to investment by foreign TNCs. Governments may also be required to cut back on health care, education, sanitation or housing spending. • Critics say that the USA and European countries exert too much influence over IMF policies.
World Bank	• The World Bank provides advice, loans and grants on a global scale. It aims to reduce poverty and to promote economic development (rather than crisis support). • In total, the World Bank distributed US$65 billion in loans and grants in 2014. For example: (1) help was given to Democratic Republic of the Congo to kick-start a stalled mega-dam project; and (2) a US$470 million loan was granted to the Philippines for a poverty reduction programme. • It has been headquartered in Washington since its establishment in 1945 at the Bretton Woods conference.	• It is argued that the World Bank has succeeded in promoting trade and development. It has helped the world avoid a return to the policies of the 1930s, when many countries put up trade barriers. This was harmful to world trade and a major contributing factor to the Great Depression of the 1930s, and the mass unemployment and hardship it brought to working people in many countries. This instability played a role in the outbreak of war in in 1939. • The World Bank can impose strict conditions on its loans and grants. Its critics describe this as 'neo-colonialism'.
New Development Bank	• In 2014, the **BRICS group** of nations announced the establishment of the New Development Bank (NDB) as an alternative to the World Bank and IMF. • In addition, China has set up the China Development Bank (CDB), which loaned more than US$110 billion to developing countries in 2010, a value that exceeded World Bank lending.	• The arrival of the NDB and CDB means that poorer nations no longer have to agree to the lending terms of the US-dominated Bretton Woods institutions (the IMF and World Bank). This can be viewed as a step towards a more democratic world order. • However, the new banks do have far less experience than the IMF and World Bank of managing global economic systems.
G8 and G20	• The G8 'Group of Eight' nations includes the USA, Japan, UK, Germany, Italy, France, Canada and Russia (conferences without Russia are called G7 meetings). Since 1975, the world's largest economies have met periodically to coordinate their response to common economic challenges. • In 2011, the G8 acted to stabilize Japan's economy after the devastating tsunami. In 2016, they met to discuss policies capable of stimulating growth in response to the global economic drag caused by China's slowdown.	• The G8 is steadily becoming less important as a forum for international decision-making. This is because several leading economies, including China, India, Brazil and Indonesia, are not G8 members. A larger group called the G20 has therefore been established which includes these leading emerging economies in addition to the G8 members. • The larger size of the G20, and the differing views of its members, sometimes weakens its ability to agree and act on issues.

4.1 Global interactions and global power

Table 4.4 Powerful global groups

Groups	Analysis	Evaluation
OECD	• Like the G20, the Organization for Economic Cooperation and Development (OECD) is another grouping of high-income nations and middle-income countries like Mexico. The OECD mission is 'to promote policies that will improve the economic and social well-being of people around the world'. • Member states have signed formal agreements on protecting the environment. They have also agreed to work together to tackle the challenge of ageing populations.	• The OECD has made good progress towards clamping down on tax evasion by TNCs. Rules to stop companies using complex tax arrangements to avoid paying corporate tax have been agreed by 31 members. They will make it harder for firms to hide money in tax havens in the future. • However, OECD economists failed utterly to predict the slowdown in the world economy which began in 2008 (see Unit 4.2, page 14). This was a huge oversight.
OPEC	• The Organization of Petroleum Exporting Countries (OPEC) is a wealthy and important global grouping of countries like Saudi Arabia and Qatar. As demand for oil has grown, OPEC nations have gained enormous wealth (Figure 4.10). • Global dependency on oil ensures OPEC countries are key political players, with real influence on the world stage.	• Several OPEC countries have suffered the destabilizing effects of civil war, insurgency or international conflict, including Kuwait, Iraq and Nigeria. There is one view that oil can actually hinder rather than help a country's development ('the oil curse' theory). • The collapse in world oil prices in 2015 left several OPEC members in need of a financial 'bail out' from the IMF.

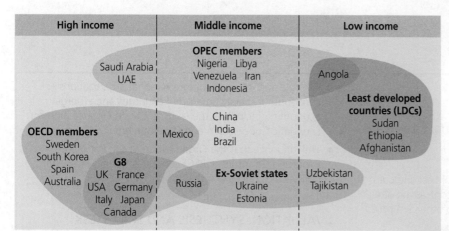

Figure 4.9 Global groupings

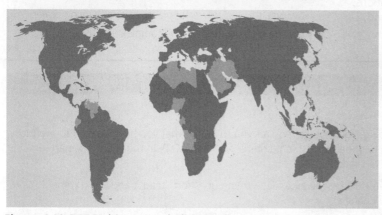

OPEC facts
- OPEC was founded in Baghdad, Iraq, in 1960, with an agreement signed by the Islamic Republic of Iran, Iraq, Kuwait, Saudi Arabia and Venezuela
- OPEC now consists of 14 member countries
- It is estimated that these 14 countries account for approximately 40 per cent of global oil production and 70 per cent of the world's 'proven' oil reserves
- When OPEC raised oil prices by 400% in 1973 it triggered an economic crisis in many countries.

Figure 4.10 OPEC's history and global influence

■ Powerful places at different scales

Some principal cities within powerful or wealthy states function as **global hubs** (Figure 4.11). These are powerful places *at both national and global scales*. Cities like New York and Mumbai attract international migrants, businesses and flows of foreign investment while also drawing in migrants and capital from other parts of the USA and India respectively. Some global hubs are megacities with over 10 million residents. Size is not a prerequisite for global influence, however. Smaller-sized global hubs which 'punch above their

> **Keyword definition**
>
> **Global hub** A settlement or state which is highly connected with other places and through which an unusually large volume of global flows are channelled.

weight' in terms of their global reach include Washington DC and Qatar's Doha, as we have already seen. In 2016, Oxford University in the UK was named as the world's leading educational institution: despite its small size, the city of Oxford is a powerful place.

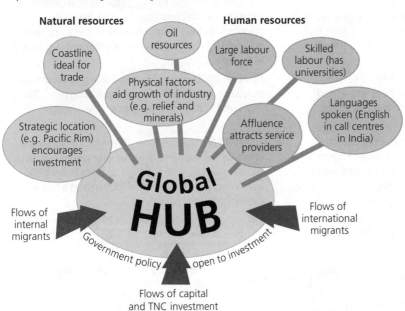

Figure 4.11 The excessive influence of global hubs often derives from an advantageous pairing of natural resources and human resources

■ KNOWLEDGE CHECKLIST

- The concept of globalization
- Attempts at quantifying globalization, for example, the Kearney index
- The global superpowers concept
- An illustration of the economic, geopolitical and cultural influence of superpowers, using two contrasting and detailed case studies (USA, China) and other examples (Qatar)
- The actions and influence of global lending institutions, including the International Monetary Fund (IMF) and China's New Development Bank (NDB)
- The role and influence of international organizations consisting of mainly developed countries and emerging economies, including: the G7/8, G20 and Organization for Economic Cooperation and Development (OECD) groups
- The influence that the Organization of the Petroleum Exporting Countries (OPEC) continues to have over energy policies and prices worldwide
- The global hub concept

EVALUATION, SYNTHESIS AND SKILLS (ESK) SUMMARY

- How wealthy and powerful places exist at varying scales
- How the globalization map is complex and dynamic over varying timescales

EXAM FOCUS

ANALYTICAL WRITING SKILLS

Once you have acquired knowledge and understanding, your course requires you to apply what you have learned to produce an analytical or explanatory piece of extended writing (under assessment criteria AO1 and AO2). In the exam, the part (a) extended writing question has a maximum score of 12 marks.

Below is a sample higher level (HL) answer to a part (a) exam-style extended writing question. Read it and the comments around it. The levels-based mark scheme is shown on page vi.

Analyse how powerful states influence global interactions in ways which benefit themselves. (12 marks)

Powerful states are countries with the means to project their influence regionally or, in the case of true superpowers, globally. Power is exercised in many ways, and different kinds of power are used by states to influence different types of global interaction. Economic power is important if you want to dominate global trade flows, for instance. Here I will be looking at economic and military power (hard power), and soft power (cultural and political influence) of different states. (1)	1 This is a useful introduction that defines key geographic terms and concepts, and also breaks them down in ways which help to structure the essay that follows.
Economic power is arguably the most important type of power as it takes money to pay for military technology and the media that are needed to project soft power globally. Large countries often have a very large GNI (though not necessarily per capita GNI). The largest states in terms of population are China, India, the USA and Indonesia, all of which are ranked	2 A well-applied account of financial flows and TNCs, which is linked well to the benefits countries gain.

highly in terms of global wealth. All of them are important global players, although the USA is still the number one global superpower according to many criteria. It has the highest nominal GNI (though China's is higher when adjusted for purchasing power parity, or PPP). Moreover one quarter of the world's 500 largest TNCs are domiciled in the USA. The USA is home to just under one-third of the world's billionaires (around 600) despite only being home to one-twentieth of the world's people. As a result of this wealth, the USA has a disproportionate influence over global economic flows and patterns of trade and investment. (2)

The USA also uses its soft power, or diplomacy, to create conditions which its businesses thrive in. US presidents have worked hard to negotiate deals with other countries which allow their TNCs to gain entry to new markets. This is done formally in meetings and also informally over dinner or a round of golf. Keeping good relations with European countries, China and India is important. So the soft power of diplomacy helps lay the groundwork for US companies to show their economic power. Some countries like the USA and UK also have a major say in how global multi-governmental organizations like the World Bank and IMF operate. They exert a lot of influence over the rules of world trade and this helps them use global interactions for their own benefit. The USA holds 17 per cent of IMF voting power and so helps decide who gets lent money and who does not. (3)

Another way in which soft power works is through the way a country's culture and values are spread across the world, the result being that other states view it more favourably. The power of TNCs and global media corporations is important again here. As well as making the USA wealthy, its TNCs have also spread American values and culture around the world, including fast food and the English language, but also things such as gender equality (because films like *Star Wars* show strong female characters). This is viewed as a positive thing by people in many other countries round the world and helps build global goodwill towards the USA. More people from other countries may want to visit the USA as tourists as a result and this helps the USA to dominate global touristic interactions too. So once again countries use global interactions to help create more economic benefits for themselves. (4)

Soft power used to be associated mainly with western nations. However, increasingly, new Asian and African powers have cultural influence too. Japan's culture has spread around the world: Sushi, Manga comics and cartoons like Pokémon are part of global culture. India's Bollywood cinema is a global export too. (5)

Finally, when we are looking at how a country uses global interactions in beneficial ways, it may be worth asking: who really benefits? A country is made up of many different individuals and stakeholders, including governments, companies and citizens, but they may not all benefit from their country's involvement in global interactions. For some people, their lives may actually get worse. Factory owners in China may have benefited from its emergence as a new superpower but not all the workers may have done. (6)

Examiner's comment

This is a well-structured (AO4) piece of extended writing, which applies (AO2) a broad range of relevant ideas, concepts and examples. The detail of the content (AO1) is good. Overall, this would reach the highest mark band.

3 Inter-governmental organizations are a relevant theme to explore when looking at superpowers.

4 Good points about tourism and cultural exchange are used to widen the response further.

5 Excellent details are provided, although this paragraph is less explicit about the benefits countries gain from their culture becoming known. Could you improve it? How does Bollywood benefit India, for instance?

6 This paragraph shows the writer is thinking very carefully about the way the question has been phrased in order to apply as much understanding to the task as he or she can. The idea of countries acting in ways that benefit 'themselves' is actually a complex proposition: not all citizens of powerful countries benefit equally from their country's success. A high level of AO2 attainment has been reached here.

Structuring an answer

The best essays use paragraphing to deliver a clear structure, which is awarded credit under criterion AO4. Read the essay and the comments and review the way that it has been carefully structured. Note that the part (a) extended writing task is not discursive and therefore does not require a formal conclusion.

4.2 Global networks and flows

In the study of global interactions, geographers conceptualize the world as consisting of networks of connected places and people. A network is an illustration or model that shows how different places are linked together by connections or flows, such as trade or tourist movements (Figure 4.12). Network mapping differs from topographical mapping by not representing real distances or scale but instead focusing on the varied level of interconnectivity for different places, or nodes, positioned on the network map.

(a) Places shown as territories on a map
(b) Places shown as nodes in a network

Figure 4.12 Places in (a) topographical and (b) network mapping

A networked world

Network flows have stimulated the imagination of writers and artists, including Chris Gray who has represented the world in the style of the iconic London Underground network map. In Gray's worldview, the divisive international borders that separate cities are no longer present and physical separation poses no obstacle to information flows between places in the internet age. The result is an isotropic surface of nodes and hubs in a borderless world, all connected by multi-coloured flow lines (Figure 4.13).

Cities viewed by the artist as being especially important or powerful are portrayed as global hubs with heightened connectivity (see Unit 4.1, page 10). Major flows of goods, services, information and people pass through or penetrate these places. Vast capital flows are routed through stock markets in global hubs such as London and Paris where investment banks and pension funds buy and sell money in different currencies. In 2013, the volume of daily foreign exchange transactions reached US$5 trillion worldwide.

PPPPSS CONCEPTS

Think about how different types of global interaction (migration, trade, internet use) may give rise to different kinds of globalizing process (cultural change, economic development, the spread of democracy).

Figure 4.13 Europe redrawn as a network (based on the London Underground design)

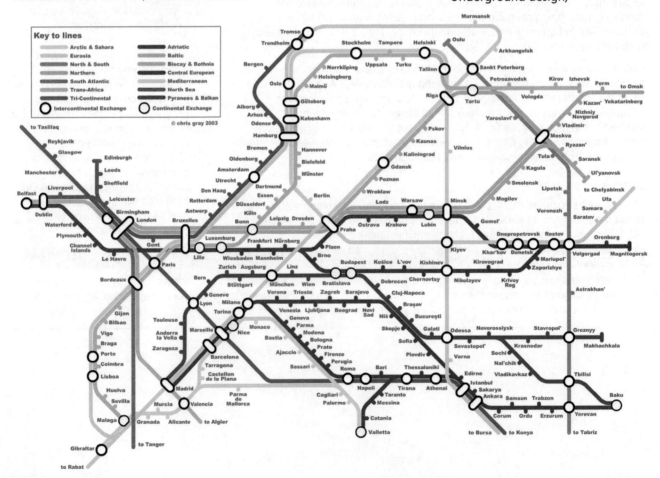

Trade in raw materials

In 2015, global gross domestic product (GDP) reached almost US$80 trillion in value. Of this, around one-third was generated by trade flows in agricultural and industrial commodities. In the past, raw material trade in food and minerals helped network states together. It remains important to global trade today, along with fossil fuel sales.

Figure 4.14 uses proportional flow lines to illustrate how raw material trade flows expanded in size between 2000 and 2010. The grey lines (the inner lines drawn for each flow) show trade volumes in 2000, while the coloured lines show the increase by 2010. The reason for this heightened activity is the rapid development of emerging economies, especially China, India and Indonesia (combined, these countries are home to 3 billion people). Rising industrial demand for materials and increasing **global middle class** consumer demand for food, gas and petrol are responsible for almost all growth in resource consumption across nearly every category shown.

> **Keyword definition**
>
> **Global middle class** Globally, the middle class are defined as people with discretionary income they can spend on consumer goods. Definitions vary: some organizations define the global middle class as people with an annual income of over US$10,000; others use a benchmark of US$10 per day income.

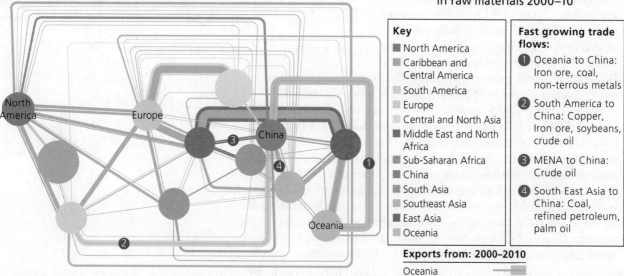

Figure 4.14 Growth in global trade in raw materials 2000–10

Trade in manufactured goods

Global flows of manufactured goods grew markedly in size during the 1990s and 2000s. The growth of trade in textiles and electronic goods was fuelled at first by low production costs in China and more recently by supply chain growth in countries where rates of pay for workers are even lower, including Bangladesh, Vietnam and Ethiopia.

Sixty years ago, the global trade pattern was very different. The majority of high-value manufactured goods were both produced and sold in North America, Europe, Japan and Australasia. Factories in these regions made use of raw materials imported from Asia, Africa and South America. Until more recently, these uneven trade flows contributed to the persistence of a global **core and periphery** system sometimes called the 'north–south divide'.

The subsequent rise of South Korea and later China (among others) as sites of innovation has transformed the global pattern of trade for manufactured goods. Korean electronics giant Samsung and China's Huawei have become major players in the production of telecommunications and home media devices. The geography of consumption for manufactured goods has altered beyond all recognition too. More than 1 billion mobile devices have been sold in India, for instance; India is also the world's fastest-growing car market. Large or fast-growing African economies including Nigeria, South Africa, Egypt and Kenya are increasingly viewed as important markets by manufacturing companies.

> **Keyword definition**
>
> **Core and periphery** In the past, the developed world 'core' regions of Europe and North America exploited the human and natural resources of the much larger and less developed 'periphery' regions of Asia, Africa and Latin America.

Trade in services

The rise of middle-class spending power in emerging economies has contributed to a rise in the trade volume of services too. In addition, regulatory and technological changes have helped shape a global market in services worth more than one trillion US dollars in 2015. Important sectors include:

- *Tourism* The value of the international tourist trade is widely believed to have doubled between 2005 and 2015 to a figure in excess of US$1 trillion (it is hard to make a precise estimate due to the many indirect benefits it creates). The number of international tourist arrivals has doubled in the same period and now exceeds 1 billion people. Much of the growth in activity has been generated by touristic movements within Asia (Figure 4.15). China now generates the highest volume of international tourism expenditure, while Europe receives more tourist arrivals than any other continent.

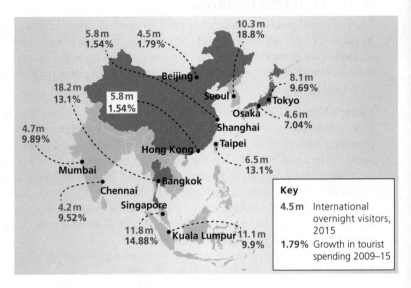

Figure 4.15 Popular Asian cities for tourism

- *Financial services and insurance* Free-market liberalization has played a major role in fostering international trade in financial services. For instance, the deregulation of the City of London in 1986 removed large amounts of 'red tape' and paved the way for London to become the world's leading global hub for financial services. Within the European Union, cross-border trade in financial services has expanded in the absence of barriers. Large banks and insurance companies are able to sell services to customers in each of the EU's member states.
- *Online media and retail* One important recent development in global trade has been the arrival of on-demand media services. Faster broadband and powerful handheld computers have allowed companies such as Amazon and Netflix to stream films and music on-demand directly to consumers. Global delivery companies such as DHL have reaped the benefits of e-commerce growth.

Some countries and world cities serve as important global hubs for particular service flows. Important stock markets are found in New York, Tokyo, Shanghai and São Paulo. Nigeria and South Korea have successful television and film industries with many viewers in neighbouring countries.

The Global Financial Crisis (GFC) and world trade

While the value of global trade has expanded enormously in recent decades, growth has slackened notably since the Global Financial Crisis (GFC) in 2007–09. The GFC originated in US and EU money markets, where sales of high-risk financial services and products triggered the failure or near-collapse of several leading banks and institutions. The resulting shockwaves undermined the entire world economy. Some countries experienced severe economic difficulty in the immediate aftermath, including Portugal, Greece and Ireland.

Several key data indicators indicate that a cyclical or longer-term downturn in world trade flows has continued to affect developed, emerging and developing economies alike since the GFC (Figure 4.16).

- International flows of trade, services and finance grew steadily between 1990 and 2007 before collapsing and stagnating. In 2016, for the fifth consecutive year, global trade did not grow. Annual cross-border capital flows of US$3 trillion are well below their 2007 peak of US$8.5 trillion.
- Oil and some natural resource prices have fallen due to the global industrial slowdown. As a result, economic growth in sub-Saharan Africa has halved, leading several countries to ask the International Monetary Fund (IMF) for help.
- A significant slowdown of emerging economies has occurred: Brazil, Russia, South Africa and Nigeria recorded minimal growth or entered recession in 2016.
- Container shipping movements have also declined. The Baltic Dry Index (BDI) – a measure of the price for shipping dry goods such as iron ore and coal – reached a record low in 2016.

China's slowdown

Another important influence on world trade is the way the Chinese economy is maturing. Despite being the world's leading economy by some measures, its growth rate has halved recently since 2007 from 14 to under 7 per cent. China was globalization's growth locomotive: slowdown of the world's largest economy has serious implications for everyone. Moreover, this is a permanent rather than cyclical change because China has entered a new and slower phase of economic development. Instead of exporting cheap mass-manufactured goods, China's leaders have shifted the country's economy towards the production of more sophisticated and higher-value consumer items for its own domestic markets. This has reduced China's demand (and therefore global demand) for natural resources, thereby ending a decade-long global commodities boom or 'super cycle'. Prices fall when markets weaken and in 2016 the prices of iron ore, aluminium, copper, gold, platinum and oil reached their lowest levels since the GFC.

> **PPPPSS CONCEPTS**
>
> Think about different future possibilities for globalization. Will the world economy pick up speed again and what could influence this?

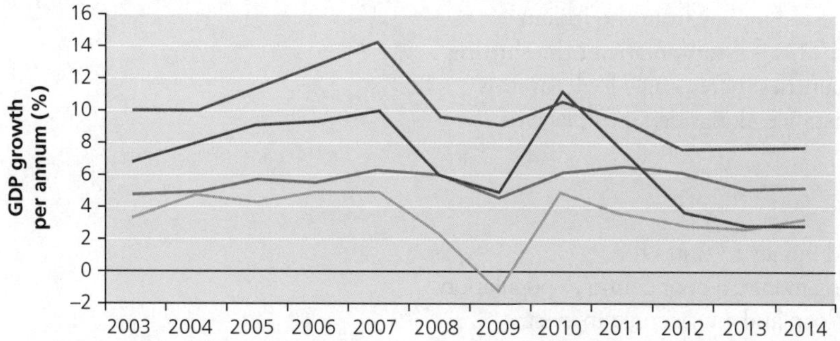

	2003	2004	2005	2006	2007	2008	2009	2010	2011	2012	2013	2014
China	10	10.1	11.3	12.7	14.2	9.6	9.2	10.5	9.3	7.7	7.7	7.4
India	6.9	8	9.1	9.4	10	6.1	5	11.2	7.7	3.8	3	5.6
Indonesia	4.8	5	5.7	5.5	6.3	6	4.6	6.2	6.5	6.2	5.2	5.2
World	3.4	4.7	4.4	5	5	2.5	-1.1	5	3.7	3	2.7	3.3

Figure 4.16 Annual economic growth for China, India and Indonesia compared with world average

International lending and debt relief

Loans can be an important financial flow for states at all levels of economic development (after the GFC, several European countries needed help from the IMF). Sums borrowed from the IMF and World Bank run into billions of dollars. In theory, borrowing brings prosperity to developing economies if the money is wisely invested in ways that earn an income both for the borrower and the lender (who charges interest). The World Bank lent Laos US$1 billion to build a dam on the Nam Theun River. The dam generates hydroelectric power. Laos can now earn US$2 billion by selling electricity to its neighbour Thailand over the next 25 years. This will be enough money to repay the loan and increase the GDP of Laos too.

IMF and WTO lending and rulings have helped many states to develop economically. For instance, the **MINT group** countries have all received large loans at some time (Table 4.5). However, after the 1970s, progressively tougher rules and conditions were attached to large-scale lending. When interest repayment rates soared after the late 1970s, Mexico threatened to default on US$80 billion of accumulated debt. Only emergency intervention from the IMF stabilized the situation.

> **Keyword definition**
>
> **MINT group** The four fast-growing economies of Mexico, Indonesia, Nigeria and Turkey.

Table 4.5 How the IMF and World Bank have supported the development of the MINT countries

Country	Development trends (2014)		IMF and World Bank support
	GDP growth rate (%)	GDP per capita (US$)	
Mexico	4	11,000	Mexico was given a vital loan of US$80 billion in 1980, without which its economy would have collapsed (taking several US banks with it).
Indonesia	6	4,000	Infrastructure was modernized in the 1970s with World Bank support; more money was lent in the 1990s after Indonesia's financial crisis.
Nigeria	7	3,000	In 2015, the World Bank offered to help Nigeria if global oil prices stay low (90 per cent of the country's foreign earnings comes from oil).
Turkey	6	11,000	In 2015, the World Bank approved a US$200 million loan to help provide finance for Turkish businesses and their supply chains.

Since then, stringent conditions have been applied to lending: **structural adjustment programmes (SAPs)** are used sometimes when states experience severe financial difficulties. Borrowing countries must agree to make concessions in return for new lending. This might involve privatizing poorly run government services, or withdrawing costly state support for inefficient industries. Critics of these concessions say they sometimes exacerbate poverty instead of solving it and further undermine the economic sovereignty of borrowing states. In critical theory, SAPs are regarded as a neo-colonial strategy used by developed countries to maintain their power and influence over how the global periphery develops (see Unit 4.1).

Critics of SAPs point to what happened in Tanzania, where poor city-dwellers were left without free water supplies after the government was impelled to privatize water supplies as a condition of its US$143 million World Bank loan in 2003. SAPs additionally generate large return money flows into the pockets of European and American financial consultancy firms (firms such as Arthur Andersen give costly advice to poor countries on any planned infrastructure improvements).

Interestingly, Tanzania has more recently borrowed money from the Indian government for its latest water supply projects. This is symptomatic of the shifting pattern of power explored in Unit 4.1: poor countries increasingly look towards new superpowers like China and India for assistance alongside, or in place of, the US-dominated Bretton Woods institutions.

Three other points to note about lending are:

- Commercial banks have sometimes loaned money to states too.
- In the event of corrupt politicians embezzling money, a poor country's population may still be required to somehow repay the loan and any accrued interest.
- At a very different scale of lending, poor people in low-income and middle-income countries borrow small sums of money called microloans from providers such as the Grameen Bank in Bangladesh (see Unit 5.1, page 48).

> **Keyword definition**
>
> **Structural adjustment programmes (SAPs)** These are money borrowing rules designed to help avoid financial mismanagement by encouraging fiscal prudence. Since the mid-1980s, the Enhanced Structural Adjustment Facility (ESAF) has provided international lending but with strict conditions attached.

■ Economic migration and remittance flows

Globalization has led to a rise in migration flows both within countries (internal migration) and between them (international migration). A record number of people migrated internationally in 2015. There are now a total of more than 250 million people living in a country they were not born in. This represents 3–4 per cent of the world's population. In fact, this percentage has not changed greatly over time despite the fact that the number of people migrating internationally has risen. This is because the total size of the world's population has grown too (between 1950 and 2015, world population grew from 4 billion to 7.3 billion).

Important changes have taken place in the *pattern* of international migration in recent years.

- As recently as the 1990s, international migration was directed mainly towards developed world destinations such as New York and Paris. Since then, world cities in developing world countries, such as Mumbai (India), Lagos (Nigeria), Dubai (UAE) and Riyadh (Saudi Arabia) have also begun to function as major global hubs for immigration (Table 4.6).
- Much international migration is relatively regionalized. In general, the largest labour flows connect neighbour countries such as the USA and Mexico, or Poland and Germany.

Table 4.6 New global hubs for migration and remittance flows

Indian workers moving to UAE	Over 2 million Indian migrants live in the United Arab Emirates, making up 30 per cent of the total population. Many live in Abu Dhabi and Dubai. An estimated US$15 billion is returned to India annually as remittances. Most migrants work in transport, construction and manufacturing industries. Around one-fifth are professionals working in service industries.
Filipino workers moving to Saudi Arabia	Around 1.5 million migrants from the Philippines have arrived in Saudi Arabia since 1973 when rising oil prices first began to bring enormous wealth to the country. Some work in construction and transport industries, others as doctors and nurses in Riyadh. US$7 billion is returned to the Philippines annually as remittances. There are reports of ill-treatment of some migrants, however.

4.2 Global networks and flows

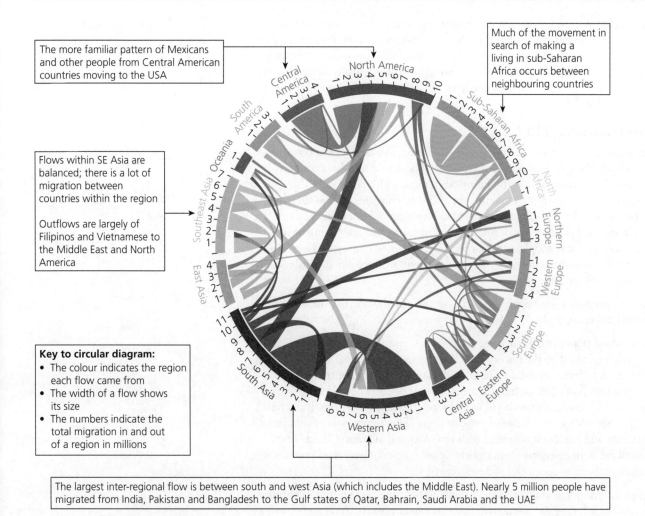

The more familiar pattern of Mexicans and other people from Central American countries moving to the USA

Much of the movement in search of making a living in sub-Saharan Africa occurs between neighbouring countries

Flows within SE Asia are balanced; there is a lot of migration between countries within the region

Outflows are largely of Filipinos and Vietnamese to the Middle East and North America

Key to circular diagram:
- The colour indicates the region each flow came from
- The width of a flow shows its size
- The numbers indicate the total migration in and out of a region in millions

The largest inter-regional flow is between south and west Asia (which includes the Middle East). Nearly 5 million people have migrated from India, Pakistan and Bangladesh to the Gulf states of Qatar, Bahrain, Saudi Arabia and the UAE

Figure 4.17 Circular flow diagram to illustrate global migration flows between 196 countries, 2005–10

■ The value of remittances

Around US$500 billion of remittances are currently sent home by migrants annually (Figure 4.18). This is three times the value of overseas development aid. In Bangladesh, the value of remittances exceeds foreign investment. Remittances can be transferred via banks or sent in the mail as cash. Unlike international aid and lending, remittances are a peer-to-peer financial flow: money travels more or less directly from one family member to another. This cross-border money flow plays a vital role in the social development of communities who have previously been excluded financially from access to education and health care.

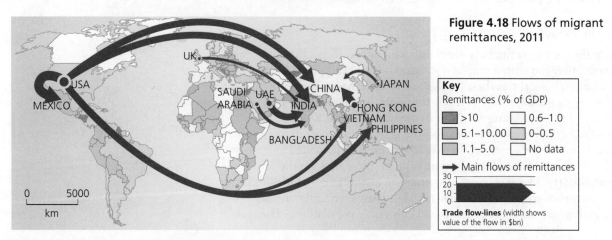

Figure 4.18 Flows of migrant remittances, 2011

Migration and remittance flows should not be seen as an inevitable consequence of globalization, however. The vast rise in trade between some countries like the UK and India has actually coincided with a decline in migration between them.

Of all global flows, the permanent movement of people still faces the greatest number of obstacles due to border controls and immigration laws. As a result, most governments have a 'pick and mix' attitude towards global flows: they embrace trade flows but attempt to resist migrant flows unless there is a special need (such as Qatar's encouragement of Indian construction workers). This theme is returned to in Units 4.3 and 5.3.

International aid flows

International aid is a gift of money, goods or services to a developing country. Unlike a loan, the gift does not need repaying. The donor may be a country, or a group of countries such as the European Union. Individuals in HICs give aid to poorer countries by making donations to charities such as Oxfam. Most international aid is targeted at specific long-term development goals for communities. Hand-pumps or lined wells can bring real improvement in quality of life for local people. On a larger scale, it can be channelled into key infrastructure improvements – for instance, German finance aided construction of Malaysia's Bakun dam.

We can see particular geographical patterns in the way that development aid is distributed internationally:

- Flows of aid from the UK are directed towards Commonwealth countries. This is partly explained by the history the UK shares with its former colonies. Until very recently, India received more UK aid than any other country (on the grounds that half a billion Indians are still very poor and need help).
- India and China now provide aid to developing African countries. India has spent US$6 billion on education projects there. The Tazara railway that links Tanzania and Zambia was funded with international aid from China. The flow of aid from emerging economies to developing countries is an important feature of the new geography of development.

However, international aid flows are dwarfed by the value of trade and investment flows. Aid alone has never raised an entire state out of poverty. Aid can be controversial also. One perspective is that it deters innovation by fostering dependency. Another view is that it neuters other potentially more important financial flows such as investment in businesses. Take the case of Zambia, for instance:

- Free provision of second-hand clothes – known locally as *salaula* – is essential for many Zambians living on around US$1.50 a day. There is often little shortage, as charities export around 80 per cent of all donated clothes to struggling nations such as this.
- However, many of Zambia's own fledgling textile industries went out of business as a result of this generosity.
- Moreover, under World Bank SAP rules attached to lending, Zambia's government was prevented from subsidizing these struggling industries.

Illegal flows

Alongside the legitimate financial flows discussed above, illegal flows of people, narcotics, counterfeit property, stolen goods and endangered wildlife link societies and places together. The United Nations has made repeated calls for states to work together to tackle transnational organized crime flows (Figure 4.19). According to UN Secretary-General Ban Ki-moon: 'Transnational criminal markets crisscross the planet, conveying drugs, arms, trafficked women, toxic waste, stolen natural resources or protected animals' parts. Hundreds of billions of dollars of dirty money flow through the world every year, distorting local economies, corrupting institutions and fuelling conflict. Transnational organized crime markets destroy, bringing disease, violence and misery to exposed regions and vulnerable populations.' The real

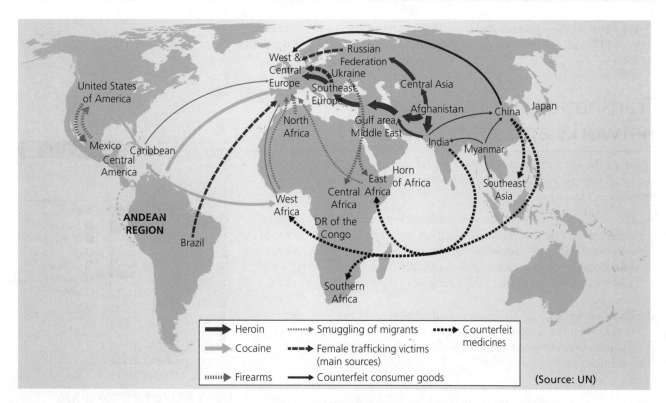

Figure 4.19 Global transnational organized crime flows (arrow width proportional to value)

size of these flows can only be estimated crudely, using police reports and anecdotal evidence.

■ Challenges and opportunities for networked and interdependent places

Over time, global flows have created networks of interconnected and interdependent places (Figure 4.20). Every country depends to some extent on the economic health of others for its own continued well-being.

- If a disaster or economic recession adversely affects a host country for migrants, then source countries may experience a drop in GDP on account of falling remittances. Some sectors of the UK economy are highly dependent on Eastern European labour; Eastern Europe, in turn, relies on migrant remittances from the UK. In 2009, during the global financial crisis, many UK building projects were cancelled. The knock-on effect was that many migrants stopped sending money; some even returned home. Estonia's economy shrank by 13 per cent.
- Social and political ties between two countries can be strengthened through migration. The arrival of a large Korean diaspora population in the USA has deepened the country's friendship with South Korea.
- Writing in the 1990s, Thomas Friedman argued that economic and political interdependency are linked. In the 'golden arches theory of conflict prevention' he asserted that two countries with McDonald's restaurants would never wage war because their economies had become interlinked. While the recent conflict between Russia and Ukraine has weakened Friedman's argument (both countries have McDonald's restaurants), it remains an idea worth exploring.
- A heightened degree of risk is also introduced to well-connected places, however. The effects of the GFC have already been outlined: its cascading effects highlighted all too well how challenges run directly alongside

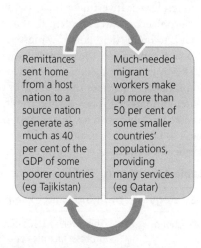

Figure 4.20 The interdependency of two networked states

PPPPSS CONCEPTS

Think about all the different ways in which global interactions can make places and societies more interdependent than in the past.

the considerable opportunities that interdependency and global network growth brings. This theme of risk exposure is explored in greater depth in Unit 6.3.

Transnational corporations: networks and strategies

TNCs are the key agents – or 'architects' – of globalization in the twenty-first century. They link states together with enormous capital flows in the form of foreign direct investment (FDI).

- The great global increase in interconnectivity and interdependency witnessed over the last 30 years owes more to FDI than any other type of monetary flow, with offshore investment by TNCs rising near-exponentially from US$20 billion in 1980 to a peak of US$1,700 billion in 2008 (before falling to US$1,200 billion by 2015 because of the recent fragility of the world economy).
- There are around 60,000 TNCs (when defined as countries with operations in more than one country). The top 100 of these own 20 per cent of world financial assets, employ 6 million people and enjoy 30 per cent of global consumer sales.

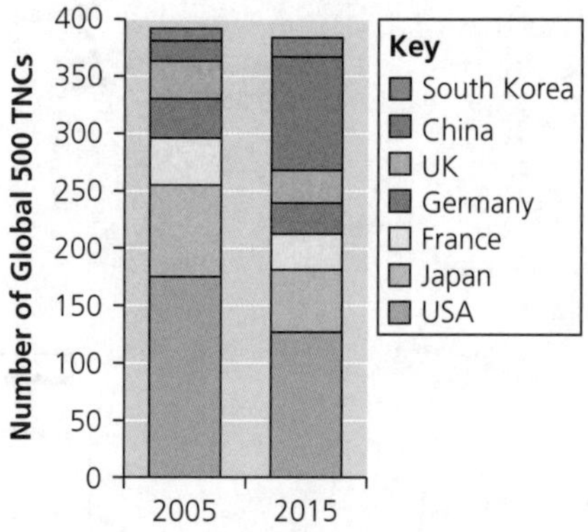

The global distribution of TNC headquarters has changed in recent years as more companies from emerging economies have appeared as major players investing in other places (Figure 4.21). Enormous outbound foreign investments are made each year by India's CFS and Infosys, as well as China's Haier and Huawei. The 2006 buy-out of Corus, previously British Steel, by Indian firm Tata was a landmark move that some commentators branded 'reverse colonialism'. Indian TNC owners – such as Tata chief Lakshmi Mittal – now feature in the 'top ten' world billionaire 'rich list'.

Figure 4.21 Fortune 'Global 500' ranking of TNCs in 2005 and 2015

■ Foreign Direct Investment strategies used by TNCs

Foreign Direct Investment (FDI) is a financial injection made by a TNC into a nation's economy. FDI consists of a range of investment strategies used by TNCs to build their global businesses, with the precise course chosen very much depending on the industrial sector the company operates within (Table 4.7). Mining and oil companies and food-producing agribusinesses do not deal in products or services that can be, or need to be, glocalized (customized for different markets). They expand into new territories primarily through mergers and acquisitions instead. In contrast, manufacturing and banking companies, such as Unilever, Samsung and Citigroup, will often modify their products and services for the diverse markets they seek to conquer.

> **Keyword definition**
>
> **Spatial division of labour** The common practice among TNCs of moving low-skilled work abroad (or 'offshore') to places where labour costs are low. Important skilled management jobs are retained at the TNC's headquarters in its country of origin.

Table 4.7 Different TNC investment strategies

Strategy	Definition and example	Evaluation
Offshoring	• This involves TNCs moving parts of their own production process (factories or offices) to other countries to reduce labour or other costs. For instance, the UK technology company Dyson moved its own manufacturing division to Malaysia in 2002. • For large media and finance companies, it can involve setting up new international offices in different territories. UK accounting and consultancy firm KPMG now has offices in over 100 states. • By offshoring, TNCs can also locate manufacturing and office services closer to the markets they will be serving. Japanese car companies like Nissan have built factories in the EU in order to serve European markets directly and avoid MGO import taxes.	• There are many benefits of creating a **spatial division of labour**, both for the TNC (whose profits rise) and the states it invests in (where jobs are created). • However, offshoring may have costs too. Job losses in the TNC's country of origin may result from the relocation of manufacturing overseas. This can attract criticism, leading to poor sales. • Also, companies may expose themselves to a range of new political and physical risks by investing in certain states. In recent years there has been a growing trend of TNCs 'reshoring' some operations in order to reduce these risks (see Unit 6.3). • Firms must also think carefully about how much decision-making power to grant to their different overseas operations.

Acquisitions	• When an international corporate merger takes place, two firms in different countries join forces to create a single entity. The publishing TNC Reed Elsevier has been a dual-listed structure since 1992, maintaining headquarters and paying corporation tax in both the UK and the Netherlands (as does Royal Dutch Shell). • When a TNC launches a takeover of a company in another country it is called a foreign acquisition. In 2010, UK chocolate-maker Cadbury was taken over by US food giant Kraft, for instance. Revenue from Cadbury's UK sales now feeds the profits of US-registered firm Mondelez International (a newly formed division of Kraft).	• In 2015, around one-third of all FDI consisted of cross-border mergers and acquisitions. Clearly, these are very important strategies which TNCs benefit from in numerous ways, including expanded markets and the opportunity to reduce costs (and therefore increase profits) through rationalization. • Changes in TNC ownership affect the geography of global financial flows. Large profit flows are redirected towards the state where the buyer is headquartered. This has major implications in turn for states and societies because of the financial losses or gains in corporation tax paid to governments.
Joint venture	• This involves two companies forming a partnership to handle business in a particular territory (but without actually merging to become a single entity). • TNCs must sometimes set up joint ventures because local investment law demands it, for instance in India. • In North India, McDonald's restaurants are part-owned by Vikram Bakshi's Connaught Plaza Restaurants. The local success of this venture owes much to glocalization strategies (see below) developed by the partnership, such as the introduction of the vegetarian McAloo Tikki Burger. It makes good business sense for the TNC to work with a local company that has a better understanding of local customer preferences.	• Setting up a joint venture reduces the risk a TNC is exposed to; it must also share the profits though. • Global financial flow patterns are affected by JVs. For McDonald's in India, only a share of the profits is transferred to the USA; the rest stays in India. • The combined expertise of a global TNC working with a local company can make the venture more successful than either stakeholder would achieve alone. The logistics of attempting to do business in many countries creates headaches for large corporations and the partner company's local knowledge may prove invaluable when trying to gain a toehold in a lucrative new emerging market (India's retail market is worth US$700 billion).
Glocalization	• TNCs sometimes invest in new product designs as part of their overseas investment strategy. Glocalization involves adapting a 'global' product to take account of geographical variations in people's taste, religion and interests. • This strategy is examined in greater depth in Unit 5.2.	• It makes business sense for some TNCs to pay attention to their customers' culture. However, not all companies need to glocalize products. For some big-name TNCs like Lego, the 'authentic' uniformity of their global brand is what generates sales. For oil companies, glocalization has little or no relevance for their industrial sector.

■ Outsourcing to other companies and places

As an alternative to investing in its own overseas operations, a TNC can choose instead to offer a working contract to another foreign company: this is called outsourcing. In recent decades, China and India have become major outsourcing destinations for manufacturing and services respectively (Table 4.8).

Outsourcing frees the TNC from the hassle of building or leasing property and employing people directly. However, it also introduces new elements of risk into the supply chain (see also Units 6.1 and 6.3). A TNC may find it harder to monitor the way in which its good or services are being made. This could jeopardize both the quality of the products and the working conditions of the people who make them. Both of these issues can affect brand reputation. The 2013 collapse of the unsafe Rana Plaza building in Dhaka, Bangladesh, led to the deaths of 1,100 textile workers. It was also deeply troubling for Wal-Mart, Matalan and many other major TNCs who regularly outsourced clothing orders to Rana Plaza.

Table 4.8 Outsourcing to China and India

	Outsourcing of manufacturing to China	**Outsourcing of services to India**
Examples	• In the 1990s, China first gained its reputation as 'the workshop of the world'. Outsourcing companies in cities like Shenzhen and Dongguan offered foreign TNCs, including household names like Bosch, Black & Decker, and Hitachi, the opportunity to have their products made at low prices using a massive pool of low-cost Chinese migrant labour. Then, it was common to hear stories of Chinese workers suffering in factory conditions similar to those of nineteenth-century Europe. • Conditions have since improved for many workers and recent strategic planning by China's government has helped some outsourcing companies move further up the manufacturing value chain. Increasingly, the manufacture of high-value products such as iPhones is outsourced to China-based companies.	• Some of India's recent economic success is attributable to the call centre services that Indian workers provide. US and UK businesses have outsourced office work there because many Indian citizens are fluent English speakers. This is a legacy of British rule, which ended in 1947. It gives India a comparative advantage when marketing call centre services to the English-speaking world. • Broadband capacity is unusually high in the city of Bangalore, which is a long-established technology hub, thanks to early investment in the 1980s by domestic companies like Infosys and foreign TNCs such as Texas Instruments. • Today, large independent Indian operators conduct contract work for all kinds of firms, from travel companies to credit card providers.
Evaluation	• Over time, the outsourcing process has resulted in a 'transfer of technology' to China. Companies like smartphone maker Xiaomi manufacture their own products instead of making goods for foreign TNCs. • However, concerns linger over the treatment of some workers in so-called 'sweatshops'. Although working conditions have since improved for many, the environment continues to suffer greatly. Dubbed 'airpocalypse' by the Western media, air pollution in industrial cities reduces Chinese life expectancy by on average five years.	• Indian outsourcing companies have become extremely profitable. Founded in 1981, Infosys had revenues of US$9 billion in 2015. It is one of the top 20 global companies for innovation. • Call centre workers earn good middle-class wages by Indian standards. However, some complain they are exploited. Their work can be highly repetitive. Business is often conducted at night – due to time zone differences between India and customer locations in the USA – sometimes in 10-hour shifts, six days a week.

■ Global production networks

The geography of many large TNCs consists of a complex web of combined offshored and outsourced operations which, in turn, serve many different worldwide markets (Figure 4.22). The resulting series of arrangements is called a **global production network (GPN)**. A TNC manages its GPN in the same way the captain of a team manages other players. As globalization has accelerated, so too have the size and density of global production networks spanning food, manufacturing, retailing, technology and financial services. Food giant Kraft and electronics firm IBM both have 30,000 suppliers providing the ingredients they need.

> **Keyword definition**
>
> **Global production network (GPN)** A chain of connected suppliers of parts and materials that contribute to the manufacturing or assembly of the consumer goods. The network serves the needs of a TNC, such as Apple or Tesco.

(a) Simple TNC spatial division of labour

A US-owned firm establishes one or two wholly-owned production bases overseas. (In the early decades of the twentieth century, large US car firms such as Ford developed 'clone' operations in countries with market potential, such as the UK, which became home to Ford's Dagenham plant in 1929)

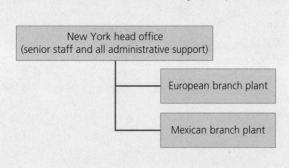

(b) Global production network (GPN)

In addition to its own offshore branch plants, this US-owned TNC is a hub company. It outsources:
• some manufacturing to a South Korean company (which, in turn, has its own supply chain)
• some administrative functions, such as call centres, to another sub-contractor

Figure 4.22 A simple TNC division of labour (a) compared with a GPN (b)

Sometimes, an extensive GPN will support the production of one single item. An amazing 2,500 different suppliers provide parts to assemble BMW's Mini car, from the engine right down to the windscreen-wipers (Figure 4.23). Some parts are outsourced from suppliers within the EU (to avoid import tariffs). In contrast, the engine comes from an offshore factory in Brazil (owned by BMW). With so many multiple upstream and downstream linkages in the supply chain, it quickly becomes impossible to pin down where a car like the Mini is really 'made'.

Figure 4.23 BMW's Mini and the global production network that supports its construction

Within supply chains, outsourcing contractors often outsource in turn to other companies. The result is a linked chain of 'tiered' suppliers. Figure 4.24 shows how Apple has outsourced construction of its iPhone to the Taiwanese company Foxconn. Two hundred suppliers (including Apple itself) provide the parts Foxconn needs. Lianjian technology is a 'fourth tier' Chinese company that delivers parts to the Korean firm Wintek, which makes the iPhone touchscreen. In 2011, it emerged that Lianjian workers had been exposed to dangerous chemicals; this generated bad publicity for Apple. Many TNCs lack direct contact with their suppliers' suppliers and only inspect working conditions at the factories of their 'first tier' suppliers. Increasingly, there is pressure for TNCs to show greater social responsibility by looking deeper into their supply chains for signs of worker exploitation (see Unit 5.1, page 50).

PPPPSS CONCEPTS

Think about the varying complexity of the supply chains and outsourcing processes used by different TNCs (for example: energy, mining, manufacturing, media companies).

Figure 4.24 A small part of the tiered supply chain for the iPhone

CASE STUDY

Fender Musical Instruments

Fender Musical Instruments Corporation has built electric guitars for 60 years, evolving over time to become a major TNC with a complex GPN. Fender has adopted a geographical location strategy that involves manufacturing very differently priced guitars in contrasting world regions. High-class instruments for professional musicians are built in the USA while mass production for high street sales takes place in a growing number of lower-wage locations spread strategically throughout the world, including Mexico, China and India. With cultural globalization encouraging worldwide diffusion of Western guitar music, Fender increasingly views these emerging economies as important new markets for its products too, and not simply places where they are manufactured.

Offshoring by Fender to Mexico began in the 1980s and accelerated after 1994 when the US, Canadian and Mexican governments agreed the North American Free Trade Association (NAFTA). The rules of this trading bloc allow US manufacturers to import materials on a tariff-free basis for assembly in low-wage Mexican factories – before transporting goods back into the USA or towards overseas markets. Low labour costs in Mexico have helped Fender to maintain a strong position in the market for cheaper mass-produced guitars.

In parallel, Fender has established its Custom Shop manufacturing division, which is located in Corona, California. The workforce is headed by highly skilled and handsomely paid 'master builders' who produce smaller volumes of much more expensive instruments carrying a 'Made in the USA' logo. These are aimed at serious musicians and collectors. This type of work is called 'post-Fordist' manufacturing (Figure 4.25). The master builders also make replicas of specific old guitars owned by famous musicians. They create purposely a 'distressed' or antique look (buyers enjoy the look of an old, scratched and discoloured electric guitar in the same way that many people like to buy new jeans with a vintage look).

Fender first entered Asian markets by establishing a joint venture in Japan back in the 1980s. Even greater success is expected in Asia in the future as new global consumer markets emerge. Cultural globalization is bringing Western rock music to India, China and Indonesia: more young people in these states now want a Fender electric guitar.

Factors encouraging lower-cost Fordist mass production in Mexico

- Cheaper wages than in USA
- NAFTA rules result in low import and export duties
- Standard range of products supplies mass market in the USA and overseas
- Low costs allow Fender to compete with low-wage Asian competitors
- Standard model assembled in Fender's Mexican factory by assembly-line workers

Factors encouraging high-cost post-Fordist production in the USA

- Skilled labour supply for the Custom Shop
- Important 'made in USA' branding for premium products
- Proximity to affluent buyers and collectors
- Californian base allows professional musicians to visit the master builders
- Custom-built by master builders in California and given a 'vintage' finish

Figure 4.25 Two brand new Fender guitars produced in different countries

CASE STUDY

Facebook

Facebook is a leading media TNC and an important architect of social and cultural globalization. In recent years, its social network has ballooned in size and influence: in 2016, there were more than 1.6 billion Facebook users (Figure 4.26). Facebook services allow users to build a global network of personal contacts, making the company a major contributor to the shrinking world effect.

- Facebook has grown its global market share in part through the acquisition of more than 50 other companies since 2005, including Instagram and WhatsApp.

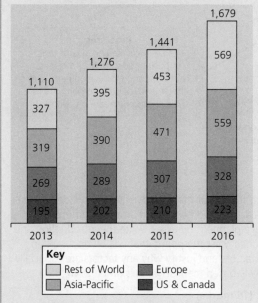

Figure 4.26 Facebook user growth, 2013–15

- The company's explosive growth has been aided by innovation from other technology companies whose ever-improving smartphone designs have brought social networks to hundreds of millions of new consumers in just a few short years.

To manage its enormous global operations, Facebook has established over 70 regional offices: in 2016, there were 24 in North America, 18 in Europe, the Middle East and Africa; 16 in Asia-Pacific and 4 in Latin America. (These can be viewed at https://www.facebook.com/careers/locations/.) Facebook FDI into other states takes two forms:

1 *Regional offices where sales strategies and advertising campaigns are developed.* There are major offices in London and Singapore, for instance. Its Indian operations are especially important: India is the largest market for Facebook outside the USA with close to 150 million monthly active users.

2 *Regional data storage facilities called server farms.* These are storage facilities that are filled with cupboard-sized racks of computer servers (giant hard drives) that store and move data such as photos, films and music. Facebook's US$760 million major data centre in Luleå is located in Sweden's coldest region. The low temperatures reduce the cost of cooling the tens of thousands of hard drives installed at the 30,000 square metre facility (the size of 11 football pitches). A flat, glaciated valley floor at Luleå provides plenty of room for future expansion, making this an interesting case of physical geography influencing FDI flows. Luleå also provides Facebook with a source of renewable energy: hydroelectric power. The TNC has been working with Greenpeace recently to develop a more sustainable business model.

■ KNOWLEDGE CHECKLIST

- Major contemporary global networks, flows and patterns
- Global trade in materials, manufactured goods and services
- The Global Financial Crisis and world trade
- Flows of international aid, loans and debt relief
- International remittances from economic migrants
- The existence of illegal flows, such as trafficked people, counterfeit goods and narcotics
- The importance of Foreign Direct Investment (FDI) by transnational corporations (TNCs) and the strategies used including offshoring, mergers, acquisitions and joint ventures
- The role of outsourcing for global businesses
- The growth of global production networks
- Two contrasting case studies of TNCs and their global strategies and supply chains (Fender, Facebook)

EVALUATION, SYNTHESIS AND SKILLS (ESK) SUMMARY

- How different flows vary in their relative importance
- How global flows and interactions are represented graphically using flow lines

EXAM FOCUS

MIND-MAPPING USING THE GEOGRAPHY CONCEPTS

The course Geography Concepts were introduced on page vii and also feature throughout Units 4.1 and 4.2.

Use ideas from Unit 4.2 (and, if relevant, Unit 4.1) to add extra detail to the mind map below (based around the Geography Concepts) in order to consolidate your understanding of global networks and flows.

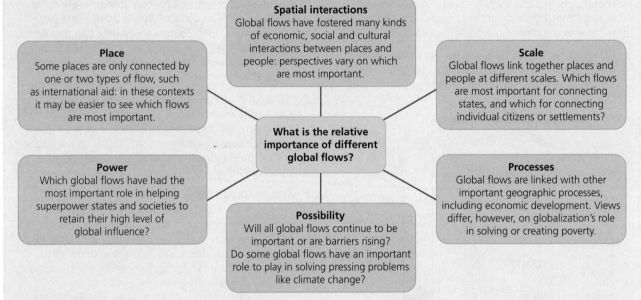

Planning an essay

Below is a sample part (b) exam-style question. Use some of the information from your mind map, or your own ideas, to produce a plan for this question. Aim for five or six paragraphs of content; each should be themed around a different key point you want to make, or a particular concept or case study. Once you have planned the essay, write the introduction and conclusion for the essay. The introduction should define any key terms and ideally list the points that will be discussed in the essay. The conclusion could evaluate the relative importance of the key factors and justify why any factors are especially important for particular geographic contexts, scales or perspectives. The levels-based mark scheme is on page vii.

Examine the relative importance of different global financial flows. (16 marks)

4.3 Human and physical influences on global interactions

Revised

Network flows rely on technology to operate. They also depend on the political will of people and their governments to 1) allow interactions with other places and 2) find the financial resources needed to overcome physical obstacles to transport and telecommunications.

Manuel Castells wrote in *Rise of the Network Society* (1996) about how technology and political decision making have allowed the global economy to function in real time as a single unit. Castells analysed how key players, such as the major TNCs, constantly scan the world for potential profit, connecting together skilled populations and affluent markets in the process. At the same time, unskilled labour markets, be they Burkina Faso or the US ghettos of Baltimore, are effectively left 'switched off' from economic globalization. A fourth world of switched-off places emerges from this network analysis: a world of social exclusion, whose inhabitants – be they subsistence farmers in Chad or homeless Parisians – can be found living in any continent or city.

This unit is mindful of the individuals and societies who remain disconnected from global interactions and the reasons for this. Personal and geographical exclusion from global networks and flows occurs for many reasons: the world we occupy is far from being borderless.

PPPPSS CONCEPTS

Think about all the possible reasons why some places are poorly connected to global networks.

Political influences on global interactions

Political factors matter greatly. Unit 4.1 explained how powerful states and organizations such as the IMF have shaped the global economy in ways that foster the kinds of free trade, foreign investment and other global flows which Unit 4.2 examined. The **neoliberal** policies that the IMF promotes are informed by a political philosophy, which the USA and other powerful states have promoted both at home and abroad since the 1980s. Also known as free-market liberalization, this governance model is associated with the policies of US President Ronald Reagan and Margaret Thatcher's UK government during the 1980s. Essentially, they followed two simple beliefs. Firstly, government intervention in markets and financial flows – at both national and global scales – impedes economic development. Secondly, as overall wealth increases, **trickle-down** will take place from the richest members of society to the poorest.

In practice, this has meant that states borrowing money from the IMF and World Bank have been obliged to remove obstacles to foreign investment. Many developing countries have been asked to privatize public services such as water or railways in order to secure World Bank or IMF lending, thereby providing foreign TNCs with the opportunity to make new acquisitions. In doing so, they adopt a political model that most developed countries have now followed for decades. Many infrastructure assets which the UK Government used to own have been privatized and sold overseas – for instance, French company Keolis owns a large stake in southern England's railway network and the UK's EDF energy company is owned by Électricité de France.

The process of globalization has been enabled by a growing political consensus that barriers to the movement of goods, services, investment and acquisitions should be removed. Many examples can, however, be found of national governments attempting to limit their spatial interactions with other states and foreign stakeholders. On the face of it, this may seem surprising, given that global flows supposedly stimulate economic growth. However, global flows are viewed sometimes from a different perspective as threatening the well-being and sovereignty of states (Figure 4.27).

> **Keyword definitions**
> **Neoliberal** A philosophy to managing economies and societies that takes the view that government interference should be kept to a minimum and that problems are best left for market forces to solve.
>
> **Trickle-down** The positive impacts on peripheral regions (and poorer people) caused by the creation of wealth in core regions (and among richer people).

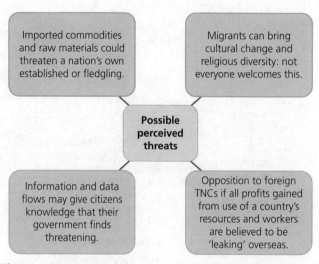

Figure 4.27 Perceived threats associated with global flows

■ Multi-governmental organizations (MGOs)

Most of the world's state governments have signed one or more political agreements to become members of **trade blocs** (Figure 4.28). These MGO agreements allow state boundaries to be crossed freely by flows of goods and money. Within a trade bloc, free trade between neighbours or more distant allies is encouraged by the removal of internal **tariffs**. This brings numerous benefits for businesses:

- By removing barriers to intra-community trade, markets for firms grow. For instance, when 10 nations including Poland joined the EU in 2004, German supermarket firm Lidl gained access to 75 million new customers.
- According to economic theories, firms possessing a comparative advantage in the production of a particular product or service will prosper in a trade bloc. French wine-makers, thanks to their advantageous climate and soil, produce a superior product that is widely consumed throughout a tariff-free Europe.
- An enlarged market increases demand, raising the volume of production, and thereby lowering manufacturing costs per unit. An improved economy of scale results, meaning products can be sold more cheaply and sales rise even further for the most successful firms.
- Smaller national firms within a trade bloc can merge to form TNCs, making their operations more cost effective. Airtel Africa is a mobile phone company headquartered in Kenya whose expansion into 17 African states has been helped by the existence of the COMESA trade blocs.

> **Keyword definitions**
> **Trade blocs** Voluntary international organizations that exist for trading purposes, bringing greater economic strength and security to the nations that join.
>
> **Tariffs** The taxes that are paid when importing or exporting goods and services between countries.

- Trade bloc members may also agree a common external tariff and quotas for foreign imports. In 2006, the EU blocked imports of underwear from Chinese manufacturers on the basis that the annual quota had been exceeded, jeopardizing sales of EU clothing makers (the media dubbed the incident 'bra wars'). More recently, EU nations have been taking steps to impose additional tariffs and quotas on Chinese steel in order to protect native European steelmakers.

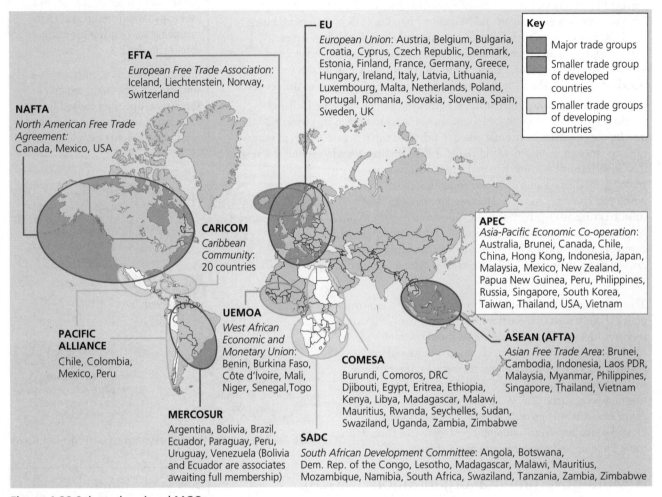

Figure 4.28 Selected regional MGOs

MGOs play a vital role in helping TNCs increase their scale of operations. Take the case of the European Union. Over time, Unilever, originally a small Dutch margarine manufacturer, has grown into an enormous conglomerate by acquiring more than 400 brands spanning food, drink and household items. Many of its acquisitions are European companies: the EU has made cross-border logistics so much easier for Unilever. Similarly, Sweden's Ikea has expanded its network of stores globally but across the EU in particular, with more than 100 mega-sized stores installed across 24 European states (while sourcing its parts and products from many of the same countries).

However, the EU is far from being completely borderless: in reality, a multitude of legal and economic obstacles to trade and investment still exist. The French and Italian governments monitor, and potentially halt, unwanted foreign takeovers in sectors deemed 'strategically important', such as energy, defence, telecoms and food.

Some MGOs are composed of states at varying levels of economic development. Unit 4.2's case study of Fender Instruments illustrated how Mexico and the USA are both part of NAFTA (North American Free Trade Agreement). This makes sense because Mexico is an emerging economy with a cheap labour force, while the USA has management and research expertise.

This allows American TNCs, such as General Electric and Nike, to optimize their use of both nations' human resources. Goods are manufactured cheaply in branch plants in Mexico (called *Maquiladoras*) but are designed and marketed by white-collar staff in the USA.

Politicians must constantly re-weigh the real or perceived benefits of free trade and MGO membership against possible costs. The sheer complexity of cross-border investment and flows complicates this task greatly. Inevitably, perspectives on the wisdom of MGO membership differ among political parties and the general public. Views are coloured by personal experiences too: some US citizens blame NAFTA for their own unemployment. Donald Trump's 2016 US presidential campaign took the same view. Trump's call for a border wall with Mexico was well-received by many fellow citizens who want stronger barriers against illegal immigration and foreign imports.

> **PPPPSS CONCEPTS**
>
> Think about varying perspectives on how MGO membership has affected the sovereignty and power of some states.

■ Free trade zones

One very important reason for the acceleration of global interactions in recent decades has been changing government attitudes in regions outside of Europe and North America. Notably, Asia's three most populated countries – China, India and Indonesia – have all embraced global markets as a means of meeting economic development goals. In every case, the establishment of **Special Economic Zones (SEZs)**, government subsidies and changing attitudes to FDI have played important roles.

> **Keyword definition**
>
> **Special Economic Zone (SEZ)**
> An industrial area, often near a coastline, where favourable conditions are created to attract foreign TNCs. These conditions include low tax rates and exemption from tariffs and export duties.

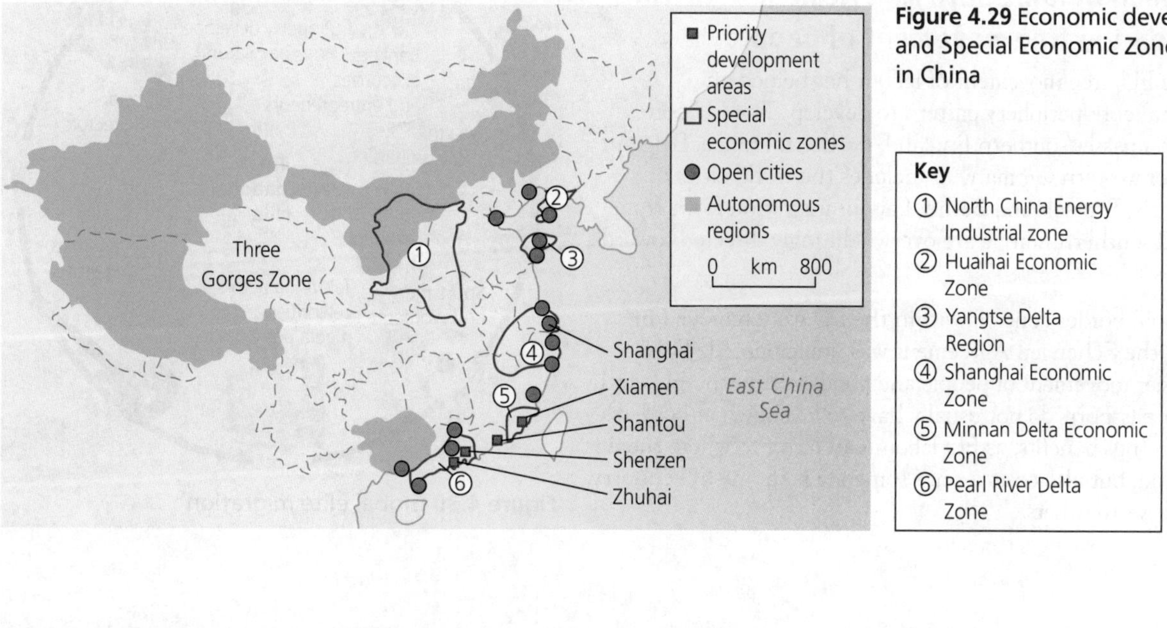

- In 1965, India was one of the first countries in Asia to recognize the effectiveness of the export zone model in promoting growth. Today, there are nearly 200 Indian SEZs.
- Coastal SEZs were crucial to China's growth in the 1980s and 1990s – many of the world's largest TNCs were quick to establish offshore branch plants or build outsourcing relationships with Chinese-owned factories in these low-tax territories (Figure 4.29). By the 1990s, 50 per cent of China's Gross Domestic Product was being generated in SEZs.
- Indonesia provides a striking example of political influences on global interactions. In the 1960s, President Suharto turned his back on communism and opened up Indonesia's markets. American and European TNCs met with Suharto's advisors and collectively built an attractive new legal and economic framework for foreign investors. The centrepiece was a new low-tax export processing zone in Jakarta – where materials could be brought into the city duty-free for pressing and manufacture prior to exporting to other markets. Overnight, Indonesia became a popular offshoring location for TNCs like Gap and Levis. World Bank lending funded the speedy modernization of it roads, power supplies and ports. However, human rights campaigners expressed concern that capital city Jakarta's export zone had become a low-tax haven for sweatshop manufacturing.

Figure 4.29 Economic development and Special Economic Zones (SEZs) in China

Key
1. North China Energy Industrial zone
2. Huaihai Economic Zone
3. Yangtse Delta Region
4. Shanghai Economic Zone
5. Minnan Delta Economic Zone
6. Pearl River Delta Zone

Migration control and rules

Individual nations vary enormously in terms of (1) the size of their population that is comprised of economic migrants and (2) the rules which govern immigration (Table 4.9). Differences in the level of political engagement with the global economy are one major reason for this. In order for a state to become deeply integrated into global systems, its government may need to adopt liberal immigration rules. Inward investment from TNCs may depend in part on the ease with which a company can transfer senior staff into a particular nation. Many of the world's leading law firms have regional offices spanning the globe, from Singapore to Moscow. In order to maintain their global networks, these companies depend on foreign states granting their staff permission to relocate permanently to overseas offices.

Top lawyers belong to a 'global elite' of professionals and high-wealth individuals (Figure 4.30). Such people are likely to encounter fewer obstacles to international migration than lower-skilled migrants. Their talent and wealth makes them more likely to qualify for a visa or residency, especially in states where a points-based immigration system exists, such as Australia. Some elite migrants live as 'global citizens' and have multiple homes in different countries.

Table 4.9 Singapore, Japan and Australia have differing attitudes towards immigration

Singapore (liberal migration rules)	Until recently, Singapore was rated as an emerging economy. Now a developed nation, this city-state is unusual in many respects. Among its 5 million people there is great ethnic diversity due to its past as a British colonial port and subsequent transformation into the world's fourth-largest financial centre. Many global businesses and institutions have located their Asia-Pacific head offices in Singapore, including Credit Suisse and the International Baccalaureate. Many foreign workers and their families have relocated there and, as a result, Singapore has many international schools.
Japan (stricter migration rules)	Less than 2 per cent of the Japanese population is foreign or foreign-born. Despite the growing status of Japan as a major global hub from the 1960s onwards, migration rules have made it tough for newcomers to settle permanently. In 2016 Japan granted asylum to just 28 refugees. Nationality law makes the acquisition of Japanese citizenship by resident foreigners an elusive goal (the long-term pass-or-go-home test has a success rate of less than 1 per cent). Japan faces the challenge of an ageing population, however. There will be three workers per two retirees by 2060. Many people think that Japan's government will need to loosen its grip on immigration.
Australia (stricter migration rules)	While Singapore has a high percentage of foreign workers, the proportion found in Australia is lower due to a recent history of restrictive migration policies. The country currently operates a points system for economic migrants called the Migration Programme. In 2013, only 190,000 economic migrants were granted access to Australia (this figure included the dependants of skilled foreign workers already living there). The top five source countries were India, China, the UK, the Philippines and Pakistan. Until 1973, Australia's government selected migrants largely on a racial and ethnic basis. This was sometimes called the 'White Australia' policy.

European Union, South American and African rules on the free movement of people

Within the EU, free movement of labour has helped an international core-periphery pattern to develop. The EU core region encompasses southern England, northern France, Belgium and much of western Germany. It includes the world cities of London, Paris, Brussels and Berlin. Labour migration flows from eastern and southern Europe are overwhelmingly directed towards these places.

Most national border controls within the EU were removed in 1995 when the Schengen Agreement was implemented. This enables easier movement of people and goods within the EU, and means that passports do not usually have to be shown at borders. Schengen brings benefits, as EU labour can move to where there is most demand, but also costs – once someone is in one EU country, they can move to others.

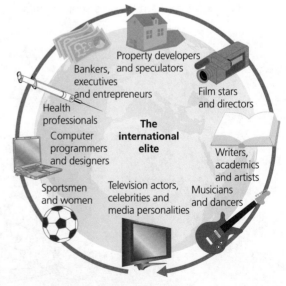

Figure 4.30 Global elite migration

Aside from the EU, what other world regions have begun to adopt free movement rules?

- South American countries have also taken steps towards this goal. Between 2004 and 2013, nearly 2 million South Americans obtained a temporary residence permit in one of nine countries implementing the agreement. After the signing of the Mercosur Residency Agreement, nationals of Argentina, Bolivia, Chile, Colombia, Peru, Uruguay and Venezuela have the right to apply for temporary residency in another member country. After two initial years of temporary residency, it is possible to convert the temporary status to permanent residency.
- The African Union has said it wants to break down borders through closer integration. In 2016, the African Union (which has 54 member states) began issuing e-passports that permit recipients to enjoy visa-free travel between member states (see Unit 6.3, page 114).

At varying times, governments try to prevent or control global flows of people, goods and information, often with mixed success. Unit 6 explores anti-immigration attitudes and laws in detail. In recent years, most EU states have witnessed the growth of popular movements opposed to this arrangement. Fears of terrorism and uncontrolled refugee movements have led some people to question the wisdom of free movement.

However, illegal flows of immigration may persist irrespective of any legal changes. Also, since 1948 the Universal Declaration of Human Rights (UDHR) has guaranteed refugees the right to seek and enjoy asylum from persecution. In 2015, large numbers of desperate refugees from Syria and poor African nations such as Somalia arrived in Europe. Some had crossed the Mediterranean in overcrowded, leaky boats with great loss of life. Many more arrived at the borders of Hungary and Serbia, having walked there. All EU states – along with most other countries – are obliged to take in genuine refugees, irrespective of whatever economic migration rules exist.

Shrinking world technologies

Revised

Improvements in both the speed and capacity of transport and ICT (information and communications technology) are frequently cast as the key 'driver' of globalization. Important developments of the last 30 years – the internet, mobile phones, low-cost airlines, to name but a few – have certainly accelerated the process. Economic transactions are now easier to complete, be they productive (manufacturers can contract and outsource physical goods from increasingly far-away places utilizing ever-faster transport networks) or consumerist (goods, information and shares can seemingly be bought anywhere, anytime, online at the touch of a button).

Heightened connectivity changes our conception of distance and potential barriers to the migration of people, goods, money and information. This perceptual change has been described as time-space convergence (Janelle, 1968) and more recently as **time-space compression** (Harvey, 1990). Janelle plotted changing travel times between Edinburgh to London and found that a two-week stagecoach journey in 1658 was ultimately superseded by air flight lasting mere hours. He concluded that different places 'approach each other in space-time': they begin to feel closer together than in the past, as each successively improved transport technology chips away more minutes and hours from the connecting journey's duration. Since the sails of ships first filled with air, human society has experienced a 'shrinking world' (Figure 4.31).

> Keyword definition
>
> **Time-space compression**
> Heightened connectivity changes our conception of time, distance and potential barriers to the migration of people, goods, money and information. As travel times fall due to new inventions, that different places approach each other in 'space-time', they begin to *feel* closer together than in the past. This is also called the **shrinking world** effect.

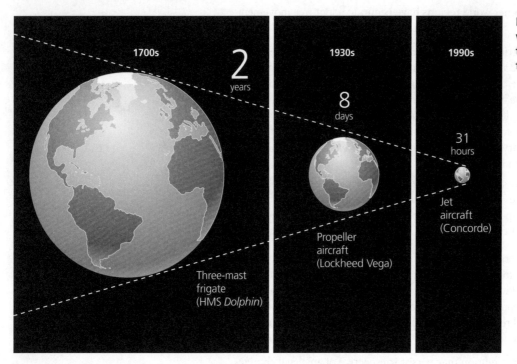

Figure 4.31 A shrinking world: the changing time taken to circumnavigate the world

In Harvey's account, time-space compression is crucial for the continued exercise of economic power. Technology has been pressed into service by global economic empire-builders. Fast trains and broadband connections are thus an *outcome* of the never-ending search for new markets and profits by TNCs. A second axis of human power – the exercise of military might and imperial ambition – continues to provide another important stimulus for new transport and communications innovation. One of the earliest **shrinking world** technologies clearly served nations' security: the practice of lighting warning fires across a chain of beacon hills dates back to ancient Greece.

This military imperative to develop new technology is also evidenced by twentieth-century design. Jet engine science was refined throughout the Second World War and during the 1950s Korean conflict. Communication satellites and major advancements in GIS/GPS owe much to the Cold War (after the Soviet Union launched Sputnik in 1957 and entered into a 'space race' with rival superpower the USA). The origin of computing also lies in wartime research and development, including the British Colossus (1943) and German Z-3 (1941) projects.

In conclusion, our analysis of technology should look at the way it has *interacted* with, and sometimes been *created by*, powerful states and companies to bring change and new possibilities. It would be oversimplifying matters to view the relationship between technology and globalization as one of simple cause and effect. Instead, it is better to discuss how they are *interrelated* with one another (Figure 4.32).

PPPPSS CONCEPTS

Think about how the shrinking world process has led to increased interactions between places. Which technologies have had the greatest impact?

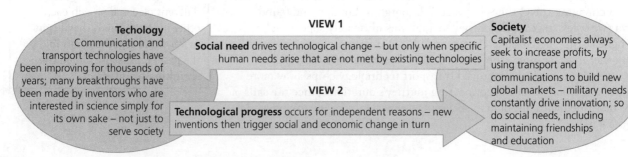

Figure 4.32 Two views on why technology changes occur

Transport developments over time

Historical studies of transport and communication are littered with innovation milestones stretching back over thousands of years. Table 4.10 shows four especially significant post-war transport innovations that have helped increase spatial interactions between places.

Table 4.10 Transport developments have led to a shrinking world

Container shipping	According to one estimate, around 600 million individual container movements take place each year. While this number may have fallen in recent years, shipping has remained the 'backbone' of the global economy since the 1950s. Everything from chicken drumsticks to patio heaters can be transported efficiently across the planet using intermodal containers. The Chinese vessel *Cosco* is 366 metres long, 48 metres wide and can carry 13,000 containers.
Lorries	Manufactured goods are the lifeblood of consumer societies. 25-metre long monster trucks keep US retail parks fully stocked with goods 365 days of the year. In recent years, the e-shopping revolution has led to the growth of 'mega-sheds': enormous strategically located warehouses where thousands of workers are constantly processing customer orders for road delivery. Amazon's Scottish mega-shed in Dunfermline is the world's largest retail distribution centre.
Air travel	The arrival of the intercontinental Boeing 747 in the 1960s made international travel more commonplace. Recent expansion of the cheap flights sector, including easyJet, has brought it to the masses in Europe: most of Europe's major cities are now interconnected via easyJet's cheap flight network (with 65 million passenger flights in 2014). The rise of the global middle class has driven the expansion of internal flights in India and China; East African Safari Air Express caters for higher earners in Kenya and its neighbours.
High-speed rail	Railways are the chief conduit linking rural and urban parts of China. Migrant workers travel in both directions along the route of the 1,500-km China–Tibet 'sky train', whose hi-tech specifications can survive the Tibetan plateau where temperatures drop to −35°C.

Global data flow patterns and trends

Alongside transport improvements, information and communications technology (ICT) has transformed the way people interact, work and consume services and entertainment. Important milestones in data transfer, storage and retrieval technologies are shown in Table 4.11. The forces behind this technological revolution are varied.

- Some important developments have been driven by military funding. The internet began life as part of a scheme funded by the US Defence Department during the Cold War. Early computer network ARPANET was designed during the 1960s as a way of linking important research computers in just a handful of different locations. Since then, connectivity between people and places has grown exponentially.
- Other crucial breakthroughs came from electronics hobbyists and university researchers: the modem device which links two computers together via conventional telephone lines (without going through a host system) was developed by two Chicago students. This vital stepping stone in the evolution of the internet came about as a result of their determination to avoid going outside during the freezing Chicago winter of 1978. (Think about which view in Figure 4.33 is supported by this.)
- Increasingly, innovation is driven by the need of TNCs to protect their market share. Samsung, Apple, Huawei and their electronics companies constantly refine their products in a competitive market that quickly becomes saturated. Sales fall once most people have purchased the latest device. In order to maintain profits, technology companies must therefore create a superior product to sell to their existing customers, ideally in a very short time-frame of just one or two years.

Table 4.11 Important elements of the growth over time of ICT

Telephone and the telegraph	• The first telegraph cables across the Atlantic in the 1860s replaced a three-week boat journey with instantaneous communication. For the first time, it became possible for people living in one part of the world to know what was happening in other places *at that same moment*. This was a truly revolutionary moment in human history.
	• The telephone, telegraph's successor, remains a core technology for communicating across distance.
Personal computers and 'the internet of things'	• The microprocessor was introduced by Silicon Valley's Intel Corporation in 1971; soon afterwards, small-scale computers began to be designed around microprocessors, including early Apple microcomputers designed by high-school drop-outs Steve Wozniak and Steve Jobs in Silicon Valley.
	• User-friendly interface technology and software were introduced first to the Apple Macintosh in 1984 and to the PC by Microsoft as Windows 1.0 in 1985.
	• Computers have evolved into laptops, tablets and small handheld devices. Small networked computers are increasingly integrated seamlessly into cars and even fridges. This new age of smart devices is sometimes called 'the internet of things'.
Broadband and fibre optics	• With the advent of broadband internet in the 1980s and 1990s, large amounts of data could be moved quickly through cyberspace. Today, the Earth's ocean floors carry enormous data flows, carried by fibre optic cables owned by national governments or TNCs such as Google. The pattern of data flow is shown in Figure 4.33.
	• More than 1 million kilometres of flexible undersea cables about the size of garden watering hoses carry all the world's emails, searches and tweets.
	• Telecoms network builders have overcome enormous challenges set by physical geography. They have slung mile upon mile of vulnerable fibre optic cable across the abyssal plains of the ocean floor. This has created new economic risks for societies: in 2006, a major submarine earthquake and landslide destroyed Taiwan's telecom link with the Philippines, disrupting TNC operations. Cyclones and tsunamis destroy cables; so too do dropped anchors.
GIS and GPS	• The first global positioning system (GPS) satellite was launched in the 1970s. There are now 24 situated 10,000 kilometres above the Earth. These satellites continuously broadcast position and time data to users throughout the world.
	• Geographic Information Systems (GIS) are a collection of software systems that can collect, manage and analyse satellite data.

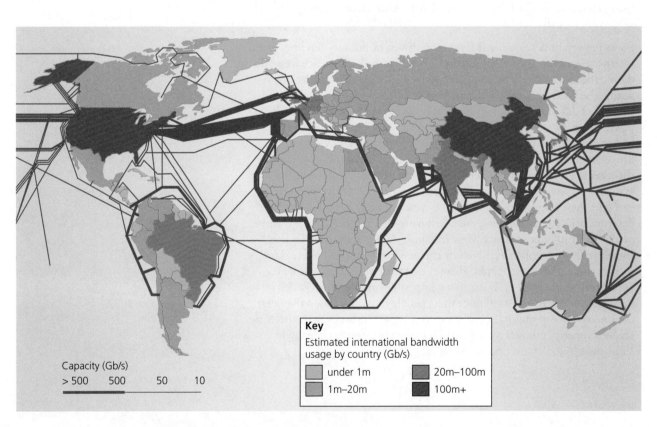

Figure 4.33 The global pattern of data flows

Patterns and trends in the use of communications infrastructure

Once available, communications infrastructure is used by citizens, business and governments in a vast array of ways that contribute to global interactions. Referring back to the framework used in Unit 4.1, we can examine how ICT has fostered different *aspects* of the globalization process:

- **Economic globalization** The offshoring and outsourcing work discussed in Unit 4.2 is facilitated by ICT. Managers of distant offices and plants can keep in touch more easily (for example, through video-conferencing). This has allowed TNCs to expand into new territories, either to make or sell their products. Each time the barcode of a Marks & Spencer food purchase is scanned in a UK store, an automatic adjustment is made to the size of the next order placed with suppliers in distant countries like Kenya. Media companies can move large data files quickly from animation studios in one country to another, thereby speeding up production time. Economic activity is supported at the personal scale too: self-employed citizens have access to crowdfunding platforms such as Kickstarter to help get their businesses started; they can also sell goods and services globally using markets like eBay or Amazon.
- **Social globalization** Migration becomes easier when people can maintain long-distance social relationships more easily than in the past using ICT. Since 2003, Skype has provided a cheap and powerful way for migrants to maintain strong links with family they have left behind. Facebook, Twitter and Snapchat work by making each individual user function as a hub at the heart of his or her very own global or more localized network of friends (Figure 4.34). Increasing numbers of people gain their education remotely by studying at a virtual school or university, or by enrolling in online MOOCs (massive open online courses). Remote health care is being provided in parts of the world where physical infrastructure is lacking. People in hard-to-reach parts of India can consult with a doctor using their mobile devices, for instance.
- **Cultural globalization** Cultural traits, such as language or music, are adopted, imitated and hybridized faster than ever before. During 2012, South Korean singer Psy clocked up over 1.8 billion online views of 'Gangnam Style', the most-watched music video of all time. Outside of the mainstream, subcultures thrive online too. Small, independent music, comic art and gaming companies can achieve an economy of scale, thanks to a digitally connected global fan base of people sharing the same minority or 'niche' cultural interest. Whether your preference is for folk music from Mali, or a specialist music subculture such as 'grindcore' or 'dubstep', you will find what you want to hear online. Cultural globalization is explored further in Unit 5.2.
- **Political globalization** The work and functioning of multi-governmental organizations (MGOs) is also enhanced by the ease with which information and publications can be disseminated. Websites for MGOs such as the European Union (EU), United Nations (UN) and World Bank (WB) contain a wealth of resources that aim to educate a global audience about issues ranging from climate change to international war crimes. Social networks are used to raise awareness about political issues and to fight for change on a global scale. Environmental charities such as Greenpeace spread their messages online. So too, however, do the militant political group Daesh (or so-called IS). Using social media and YouTube, Daesh has published horrific films of executions online. Its internet propaganda has succeeded in attracting numerous young men and women from across Europe, Asia and Africa to fight in the Middle East.

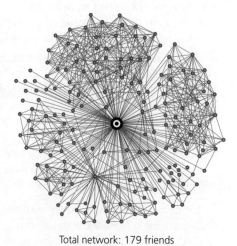

Total network: 179 friends

Figure 4.34 One person's Facebook friends visualized as a personal network

It remains the case that not everyone can participate in global interactions using ICT even in a place where it is available. Doreen Massey was a geographer who wrote critically about changing perceptions of place in a technologically advancing world. She argued that time-space compression is socially differentiated: not everyone experiences the sense of a shrinking world to anything like the same extent because of income differences. Billions of people still cannot afford the cost of a smartphone and broadband subscription. Political factors also play a role in the persistence of a digital divide between 'switched-on' and 'switched-off' individuals and societies (these are explored in Unit 5.3).

The mobile phone revolution and electronic banking in developing countries

Lack of communications infrastructure used to be a big obstacle to economic growth for developing countries. Now, however, mobile phones are changing lives for the better by connecting people and places. The scale and pace of change is extraordinary.

- In 2005, 6 per cent of all Africans owned a mobile phone. By 2015 this had risen ten-fold to 60 per cent due to falling prices and the growth of provider companies such as Kenya's Safaricom. Only 10 per cent of Africa's population live in areas where no mobile service is available.
- Rising uptake in Asia (in India, over 1 billion people are mobile subscribers) means there are now more mobile phones than people on the planet (Figure 4.35).

In 2007, Kenya's Safaricom launched M-Pesa, a simple mobile phone service that allows credit to be directly transferred between phone users. The mobile banking system M-Pesa has spread throughout East Africa and India to become an international phenomenon. In Kenya, over 10 million people either access their bank accounts or send peer-to-peer money payments using their mobile phones. This has revolutionized life for local individuals and businesses:

- The equivalent of around one-half of the country's GDP is now sent through the M-Pesa system annually.
- People in towns and cities use mobiles to make payments for utility bills and school fees.
- In rural areas, fishermen and farmers use mobiles to check market prices before selling produce.
- Women in rural areas are able to secure microloans from development banks by using their M-Pesa bills as proof that they have a good credit record. This new ability to borrow is playing a vital role in lifting rural families out of poverty.

As the proportion of mobile users with internet access (rather than simply text messaging) grows, the population is gaining access to apps for health care, education, finance, agriculture, retail and government services. In Kenya and neighbouring countries, there is already an East African app for almost everything: herding cattle in Kenya (i-Cow), private security in Ghana (hei julor!) and remotely monitoring patients in Zimbabwe (Econet). In Uganda, a new mobile service (Yoza) connects people with dirty laundry to mobile washerwomen.

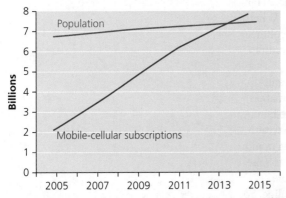

Source: ITU World Telecommunication/ICT Indicators database
Figure 4.35 The great convergence: numbers of people and mobile phones

PPPPSS CONCEPTS
Think about how small-scale interactions using mobile phones could possibly help an entire state to develop economically.

The physical environment and global interactions

Revised

What role does the physical environment play in determining how globalized places become? Common sense suggests that countries with a long coastline and an abundance of valuable **natural resources** will grow rich through trade. In reality, this hypothesis – which is sometimes called environmental determinism – is far from proven.

Natural resource availability

Some natural resources are renewable (sustainably managed forest, wind power and solar energy); other are non-renewable, including fossil fuels, metal ores and rare earths. Examples can be found of instances where natural resources have underpinned economic development and helped some states to become powerful global hubs:

> **Keyword definition**
> **Natural resources** Parts of the physical environment that are used to satisfy human needs and wants.

- Resources can be traded to provide capital or processed to make value-added manufactured goods; early-industrializing states like the UK and Germany possessed plentiful supplies of coal and metal ores.
- More recently, some oil-rich Middle Eastern states have grown enormously wealthy, including Qatar (see Unit 4.1, page 6). The vast wealth of the United Arab Emirates stems from the discovery of almost 10 per cent of the world's oil reserves beneath its sands. The leaders of Abu Dhabi – another important global hub and the second most populous city in the UAE – have spent billions of dollars investing in a range of global brands including Ferrari cars, the UK's Manchester City football club and France's Sorbonne university. Oil does not last forever and the intention is to diversify the economy in ways that turn Abu Dhabi into a culturally rich global hub, which can attract elite expatriate migrants from other countries.

However, other examples demonstrate the links between natural resources, globalization and economic growth are far from clear-cut. Known as Zaire until 1997, Democratic Republic of the Congo (DRC) is rich in natural resources. Writing in the 1980s, eminent geographer Michael Chisholm forecast a bright future for Zaire on account of its natural resource endowments. However, today DRC is ranked 176th in the Human Development Index. Life expectancy is under 50 and most people live on less than US$1.25 a day (the absolute poverty line). A mere 3 per cent of people have internet access, making it one of the least connected states in the world. So why is there a lack of correlation between resource availability and global connectedness for DRC?

Table 4.12 shows a timeline of changes affecting DRC before and after its independence in 1960. DRC's 'resource curse' is that its enormous raw material wealth has attracted numerous outsiders who eventually find local collaborators to help them loot the country's natural resources. Elsewhere in central and western Africa, natural resources have come to be associated with malingering poverty and oppressive government too. This is because of the following:

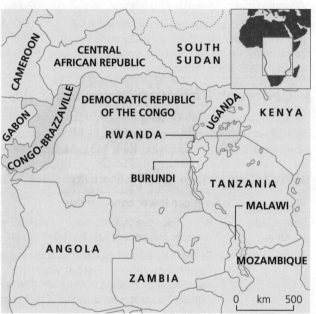

Figure 4.36 DRC shares a border with nine neighbour states, many of whose armies have invaded and looted its raw materials

- The scramble for control of natural resource revenues may result in corruption and even civil war or invasion by another sovereign state. Sierra Leone is greatly endowed with diamonds, yet is also the scene of a brutal civil war that waged throughout the 1990s. This war was, in part, related to control of the diamond trade – the 'blood diamonds'.
- Natural resource wealth does not necessarily trickle down to the poorest people.
- The discovery of natural resources may lead to other industrial enterprises being abandoned, lessening prospects for sustainable economic development and global connectivity.

Table 4.12 The link between development, globalization and natural resources in DR Congo

Geopolitical changes	Factors influencing change
Colonized by Belgium 1870–1960 Millions of Congolese were killed or worked to death by King Leopold II of Belgium.	Raw materials drew Europeans to the region. Timber, rubber, copper, cobalt, diamonds and gold made the country an important prize.
Independence as Zaire 1960–1990s After a power struggle, Joseph Mobutu took power and renamed the country. He created a difficult regime for TNCs to operate in, while amassing a US$4 billion fortune for himself. Zaire eventually defaulted on loans, resulting in a cancellation of development programmes.	Geopolitical strategies followed by Belgium and the USA during the Cold War helped put Mobutu in charge (he was favoured because of his generally pro-Western stance). However, Mobutu's family siphoned off profits from the sale of the country's natural resources for themselves, leaving ordinary people in great debt.
Regime change and conflict 1990s–2005 Mobutu was removed from power when neighbour states of Uganda and Rwanda assisted a rebel leader, Laurent Kabila, to become president in 1997 (he renamed Zaire as DRC). Later, they turned on him only to discover other neighbours (Zimbabwe, Angola and Namibia) had sent troops to 'help' Kabila. The six-nation war claimed 5 million lives.	Cross-border ethnic ties between some Rwandans and some Congolese are a legacy of colonialism and an important reason why the conflict came to involve more than one nation. Raw materials once again proved to be a source of prolonged conflict. The United Nations believes many occupying forces were motivated in part by a desire to grab DRC's resources.
Attempted conflict resolution 2005–present Global organizations now take greater interest in DRC: peacekeepers are trying to bring stability in the biggest operation the UN has ever mounted; the World Bank has approved US$8 billion debt relief.	Displaced refugees and traumatized civilians have yet to be rehabilitated. Militia groups still operate in the east of the country. However, the global internet campaign Kony 2012 put pressure on central African governments to tackle these armed terror groups.

Geographic isolation at varying scales

Some countries are relatively geographically isolated: few of the world's landlocked countries and remote islands have become highly networked global hubs. Does location really affect connectivity? Table 4.13 evaluates some of the evidence.

> **PPPPSS CONCEPTS**
>
> Think critically about the relationship between the physical environment and different possibilities for trade and globalization.

Table 4.13 How location affects connectivity

Factor	How it can lower connectivity	Evaluation
Landlocked countries	• With a few exceptions, the world's 45 landlocked countries are poor and have low levels of trade. • Of the 15 lowest-ranking countries in the Human Development Index (see Unit 5.1), eight have no coastline. All of these are in Africa and their per capita GDP is 40 per cent lower than that of their maritime neighbours. • Their most obvious handicap is in moving goods to and from ports. • Some people think landlocked countries have not enjoyed the historic benefits of global flows of migration, ideas and new cultural ideas. Global flows that brought innovation to maritime countries may have largely bypassed landlocked ones.	• Some landlocked countries have developed economically and become important global hubs; being landlocked is not always a barrier to global interactions. MGO trade agreements may be more important than location. • Landlocked Switzerland is a major global finance centre and the headquarters of many TNCs including UBS and Credit Suisse. Botswana is a middle-income landlocked country which exports diamonds using global air networks. • Maritime states contain isolated rural areas whose populations are relatively unconnected with the rest of the world either by choice or constraint (such as the Amish community in USA, or Amazonian tribes in Brazil).
Remote island nations	• Saint Helena is a tiny volcanic island with 4,200 inhabitants in the middle of the south Atlantic (Figure 4.37). The only way to get there is a five-day journey on a Royal Mail ship from Cape Town. A new airport failed to bring new flows of tourists to St Helena because strong winds make landing there too dangerous; there are still just 150 tourist beds there. • Until the mid-1900s, the physical isolation of Iceland in the North Atlantic Ocean ensured that it remained relatively disconnected from the rest of the world. Icelanders retained a strong sense of cultural unity, exhibited through ancient rituals like eating 'rotten shark' together.	• In recent decades Iceland has overcome its isolation to become far more connected. The island's population of 320,000 people is swelled by an average of 30,000 tourists each day of the year, who flock to Iceland to see its geography. Many young Icelanders are avid consumers and creators of global culture. • However, connectivity exposed Iceland to risk too: some globally connected Icelandic businesses suffered heavy losses in the GFC. • Other remote but well-connected islands include the Hawaiian archipelago (part of the USA) and the Galápagos Islands (a biodiversity visitor hotspot).

4.3 Human and physical influences on global interactions

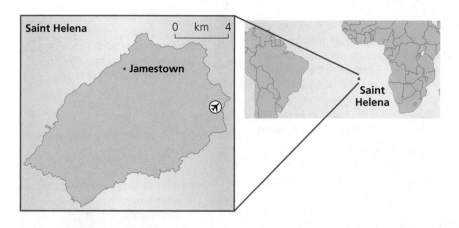

Figure 4.37 Saint Helena is one of the world's more disconnected remote islands

■ KNOWLEDGE CHECKLIST

- The political factors that affect global interactions
- Multi-governmental organizations (such as the EU and NAFTA)
- The importance of Special Economic Zones (such as are found in China, Indonesia or elsewhere)
- Economic migration controls and rules (including the EU Schengen Agreement)
- The shrinking world concept
- Transport developments, including container shipping, high-speed rail and long-haul flights
- Changing global data flow patterns and trends
- Patterns and trends in communication infrastructure and use, such as undersea fibre optic cables
- The mobile phone and internet technology revolution in developing countries
- The positive and negative influence of the physical environment on global interactions
- Natural resource availability (including the role of fossil fuels)
- The potential (but by no means absolute) limiting effect of geographic isolation, at varying scales

EVALUATION, SYNTHESIS AND SKILLS (ESK) SUMMARY

- How different globalizing factors and processes are interlinked and interact in complex ways
- How globalization has varied in its speed and extent over time

EXAM FOCUS

SYNTHESIS AND EVALUATION WRITING SKILLS

In addition to acquiring knowledge and understanding, your course requires you to **synthesize** and **evaluate** your information (under assessment criterion AO3). Your ability to carry out synthesis and evaluation is assessed in the exam's part (b) essay question, which has a maximum score of 16 marks.

- **Synthesis** means linking together different relevant themes you have studied in order to carry out your discussion or examination. Good answers will choose selectively the themes, concepts and ideas which help to answer the question most effectively. The answer to the question below synthesizes several different themes taken from this unit. As you progress through this book, you will have more different themes to link together. Ultimately, you should aim to use information from several different units when writing a part (b) answer. When you reach the end of Unit 5.3 you will find another example of a part (b) essay answer that does this very well.

- **Evaluation** means weighing up different factors, views or impacts and arriving at an overall judgement. The task may involve discussing a viewpoint or examining a particular pattern, trend or phenomenon, such as the geography of power. A good evaluation will almost always make explicit use of specialized geographical concepts such as scale, place or power; or will approach the debate or issue from a variety of stakeholder perspectives.

Below is a sample higher level (HL) answer to a part (b) exam-style question. Read it and the comments around it. The level-based mark scheme is on page vii.

Examine the interactions between technology, transnational corporations (TNCs) and the growth of globalization. (16 marks)

Globalization is the way the world's different states and societies have increasingly joined together to create a single economic system and shared culture. Technology has played an important role in this, particularly with the arrival of the internet and social networks. TNCs are companies whose operations are spread across many countries and places. They have been described as architects of globalization. For instance, big manufacturing firms like General Motors and Ford have played an important role in economic globalization, with companies like Apple also helping social globalization. This essay will examine how all these different factors and issues are interrelated. (1)

1 This is a useful introduction which defines two geographic terms and concepts, and breaks them down in ways which already begin to show understanding of what 'interactions' are.

The shrinking world effect describes how places feel closer together than they used to due to time-space compression, a process which has been happening for centuries. (2) In the 1800s, railways and steam ships helped connect countries and continents, for instance the first telegraph line between Europe and the USA in the 1870s can be viewed as early globalization. Since 1945, container vessels have played a key role moving goods from China and other emerging economies to global markets. Jet planes and cheap airlines allow millions of people to participate in global tourist flows. This links places financially and can bring cultural change too.

It should be pointed out that some interactions between technology and globalization result from the deliberate actions of powerful countries and international organizations. Early telegraph and telephone lines were paid for by European governments who wanted to build global empires. Railways and motorways have been built for strategic reasons too. Most recently, the internet grew out of US defence department research during the Cold War. The idea was to link places together to survive a military attack. (3)

Turning now to TNCs, they are also a major factor for globalization. Large TNCs, including McDonald's, Nike and banks and services like Barclays and HSBC, have built extensive global production networks. They have used foreign direct investment in a range of ways both to build low-cost factories and offices in some countries and to develop other countries as new markets. The biggest companies use a range of strategies, including outsourcing, mergers and acquisitions, and glocalization to spread themselves as widely as possible. (4) McDonald's has entered over 100 countries using glocalization. This means adapting its products to fit the tastes of different societies. Media and technology companies like Facebook and Disney have a very large influence on a global scale. They have encouraged the spread of the English language and also Western traditions like Christmas, Halloween and Valentine's Day (although some places have tried to resist this, e.g. Valentine's Day was banned in rural Pakistan).

Taking this approach further, there are many different types of TNC operating in different sectors. We should not assume that manufacturing and technology companies are the only ones who play an important role in globalization. (5) There are also food, mining and energy companies to consider. Some of the world's agri-businesses, like Del Monte, have globalized their operations. Cash crops are grown in many countries to feed rising consumer demand in developed countries and emerging economies. Part of globalization involves the spread of middle-class lifestyles – billions of people now have a meat and dairy diet. The media, spread through technology, are partly responsible for this because advertising by TNCs like McDonald's creates new aspirations. This is the reason why diets are changing across Asia and, increasingly, Africa too.

It is also important to realize that different types of technology have a role supporting different aspects of globalization. (6) I have already written about container ships supporting trade, and the internet spreading culture. There are other links and interactions too. Skype and electronic remittances through the internet support the record amount of economic migration taking place today. Over 250 million people live in countries they were not born in. Being able to communicate easily with family at home clearly plays a role in supporting the global flow of people.

The USA is home to Koreans, Filipinos and Europeans who use shrinking world technology to stay in contact with their families. Finally, the internet even supports political globalization campaigning organizations like Greenpeace and Amnesty fundraise and raise awareness globally. Many young people think of themselves as global citizens because of their involvements in campaigns like Kony 2012 and #bringbackourgirls. Sadly, political organizations like ISIS are able to spread globally too.

In conclusion, there are many ways in which technology, TNCs and globalization are interrelated and interact. TNCs like Apple, Google and Microsoft have driven technical innovation in the era of globalization. In developed countries, where a market can quickly become 'saturated', the research focus of companies tends to be expensive 'improved' products – accompanied by adverts that can convince existing customers to throw away the fully working smartphone that they already possess! So whenever their profits are falling, there is an impetus for many firms to heavily invest in research and development of new products. The technologies we use are developed by companies, governments

---

2 A well-applied 'cause-and-effect' account of the shrinking world technology and how it has driven globalization.

3 Some good extended points about the power of governments to help develop new technologies in order to promote globalization.

4 A well-applied 'cause-and-effect' account of TNCs and how they have helped accelerate globalization.

5 Some good extended points about different types of TNCs and varying aspects of globalization. The command word 'examine' is usually defined as: 'Considering an argument or concept in a way that uncovers the assumptions and interrelationships of the issue'. Here, the writer does this by thinking critically about how the question statement might apply to different kinds of TNC.

6 This paragraph shows the writer is still thinking very carefully about possible underlying assumptions behind the essay statement. The idea of different types of technology and different kinds of TNC assisting different aspects of globalization shows a high level of AO3 achievement.

7 The final paragraph offers concluding evaluative remarks while also making the final developed point that interactions between technology, TNCs and globalization are a 'two-way street'.

and individuals who are all responding to a real or perceived social demand for even greater connectivity. Businesses want faster broadband and processing power, to keep their competitive edge. Families say they want even faster or sharper Skype, so they can telecommunicate with even greater clarity. Consumers ask for fewer limits on the amount of on-demand digital media they can download and consume. The cycle of innovation is accelerated further by corporate mergers. Microsoft recently acquired Nokia, for instance, and Google has bought Motorola. For this reason, constant innovation takes place in the fields of computing, handheld devices, software, broadband and wireless technologies – all of it driven by TNCs who are scared of losing their position as major globalization players. (7)

Examiner's comment

The synthesis and evaluation (AO3) shown here go far beyond a simple 'cause-and-effect' argument and would merit a top-level mark. One of the reasons why this essay is so successful is that it draws many multi-directional links between the ideas it examines. The structure (AO4) is very good. The recall (AO1) and application of knowledge are good, if occasionally generalized.

Content mapping

The further you progress through the course, the greater the amount of materials that are available for you to choose from as you carry out your synthesis. This answer has made use of case studies, concepts, theories and issues drawn from all of the first three units of the book. Using the book's index, try to map the content.

Unit 5 Human development and diversity

5.1 Development opportunities

Revised

Human development, like globalization, is a multi-dimensional process which can be studied at varying geographic scales. There is a large area of overlap between development studies and globalization studies:

- Both topic areas are concerned with economic disparities and the factors that can reproduce or reduce them.
- The financial flows that allow global systems to operate also transfer wealth between places in ways which can narrow or exacerbate different kinds of **development gap**.

> **Keyword definition**
>
> **Development gap** A term used to describe the polarization of the world's population into 'haves' and 'have-nots'. It is usually measured in terms of economic and social development indicators. Development gaps exist both between and within states and societies.

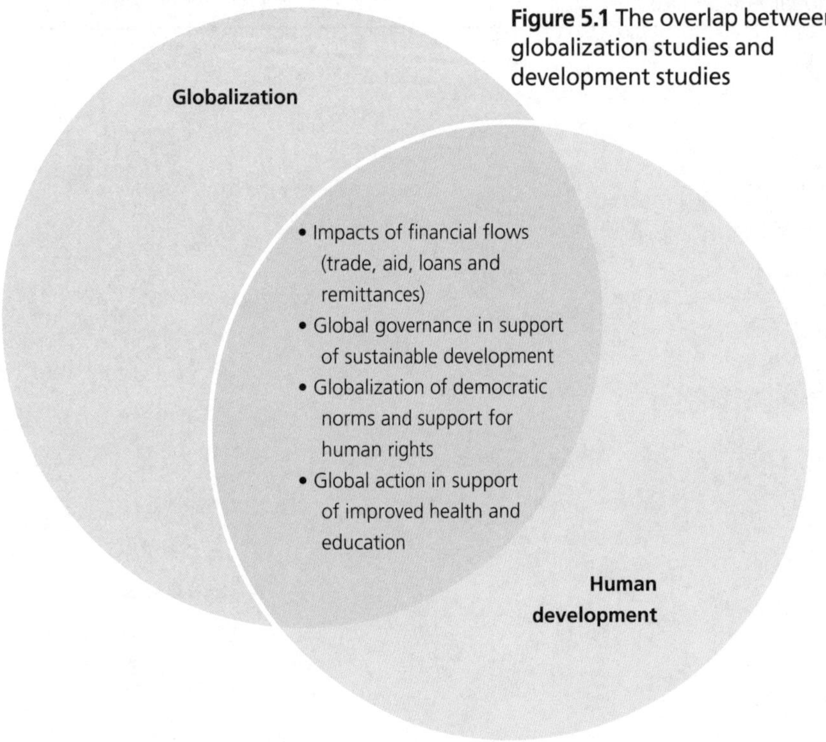

Figure 5.1 The overlap between globalization studies and development studies

Globalization

- Impacts of financial flows (trade, aid, loans and remittances)
- Global governance in support of sustainable development
- Globalization of democratic norms and support for human rights
- Global action in support of improved health and education

Human development

> **PPPPSS CONCEPTS**
>
> Think about the importance of geographical scale for the study of development. Important questions to ask are 1) do all local places within a country have the same level of development and 2) do all groups of people share the same economic and social opportunities?

Human development and ways of measuring it

Revised

Human development generally means the ways in which a country seeks to progress economically and to improve the quality of life for its inhabitants. A country's level of development is shown firstly by economic indicators of average national wealth and/or income, but encompasses social and political criteria also. Figure 5.2 shows 'the development cable'. It presents the development process as a complex series of interlinked outcomes for people and places. In summary, it shows that in an economically developed society:

- citizens enjoy health, long life and an education that meets their capacity for learning
- citizenship and human rights are more likely to be established and protected.

Figure 5.3 shows the social changes that sometimes follow when the world's poorest farmers receive a boost in earnings. It highlights how economic, social, cultural and political changes can become interlinked as part of the human development process.

5.1 Development opportunities   43

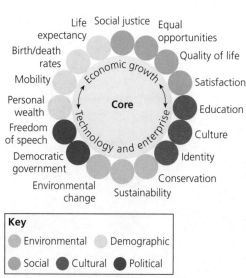

Figure 5.2 The development cable

Figure 5.3 Economic and social development linkages

■ The UN Sustainable Development Goals (SDGs)

The UN's 17 Sustainable Development Goals (SDGs) were introduced in 2015. They replace and extend the earlier Millennium Development Goals (MDGs) which were a set of targets agreed in 2000 by world leaders. Both the SDGs and earlier MDGs provide a 'roadmap' for human development by setting out priorities for action. Figure 5.4 shows the SDGs: as you can see, there is a strong emphasis on poverty reduction and the promotion of health, education and sanitation. It is hard to imagine why anyone might object to these goals. You may, however, be able to conceive of some individuals or societies being less convinced of the need for Goals 5 and 13 because of their own perspectives on these issues, which may differ from your own.

Figure 5.4 The UN Sustainable Development Goals for 2030

The SDGs were agreed on following consultations with governments and other stakeholders in many countries. The overarching aim is to end poverty, fight inequality and injustice, and tackle climate change by 2030. The SDGs also interlink with the three strategic focus areas of the UN Development Programme (UNDP):

- sustainable development
- democratic governance and peace building
- climate and disaster resilience.

The validity and reliability of human development indicators

Human development is measured in many different ways using both single and composite (combined) measures. When assessing the value of different measures it is helpful to distinguish between issues of validity and reliability.

- For a measure to be *valid*, there should be broad agreement that it has relevance. Do you agree that political corruption should be used as a measure of development, for instance? Should gender equality be included as one of the most important aspects of human development, as the SDGs have done? Should we consider a country's commitment to sustainable development policies and climate change mitigation when attempting to analyse varying levels of human development?
- To be *reliable*, a measure must use trustworthy data. Do you think all countries' income and employment data are fully accurate, for instance? Might data measurement or comparability issues mean that we should sometimes question the reliability of different countries' estimates of their fertility, mortality or literacy rates? Should estimates of illegal global flows (see Unit 4.2, page 18) be included in estimates of national income and wealth creation?

Table 5.1 explains how three important human development indicators are used and offers an evaluation of their validity and/or reliability.

Table 5.1 Evaluating different development indicators

Indicator	Explanation	Evaluation
Income per capita	• Income per capita is the mean average income of a group of people. It is calculated by taking an aggregate source of income for a country, or smaller region, and dividing it by population size to give a crude average (which can give a misleadingly high 'typical' figure if large numbers of high-earners inflate the mean). • Per capita gross domestic product (GDP) is one of the most widely used proxies for this. It is the final value of the output of goods and services inside a nation's borders (i.e. a crude estimate of the entire nation's income). Each country's annual calculation includes the value added by any foreign-owned businesses that have located their operations there. • The World Bank recently estimated global nominal GDP in 2014 at about US$78 trillion. Using this figure, can you make an estimate of global GDP per capita?	• There is near-universal agreement that it is perfectly valid to study income levels as part of an enquiry into human development. • Recording GDP reliably is not always possible though. The earnings of every citizen and business need to be accounted for, including **informal sector** work. Also, to make comparisons, each country's GDP is converted into US dollars. However, some data may become unreliable because of changes in currency exchange rates. • GDP data must be manipulated further to factor in the cost of living, known as purchasing power parity or PPP. Simply put, in a low-cost economy, where goods and services are relatively affordable, the size of its GDP should be increased and vice versa. This is why websites often provide you with two estimates of a country's GDP (Brazil had a 'nominal' GDP of US$1.9 trillion in 2015, and a 'PPP' GDP of US$3.6 trillion).
Human development index (HDI)	• The HDI is a composite measure (Figure 5.5). It ranks countries according to economic criteria (gross national income per capita, adjusted for purchasing power parity) and social criteria (life expectancy and literacy). • It was devised by the United Nations Development Programme (UNDP) and has been used in its current form since 2010. • The three 'ingredients' are processed to produce a number between 0 and 1. In 2014, Norway was ranked in first place (0.944) and Niger was ranked in last place (0.337).	• The three ingredients of the HDI – wealth, health and education – are widely regarded as valid indicators of development. All governments value wealth and health, while the education of citizens (as indicated by literacy) plays a crucial role supporting these and other goals. • Literacy and life expectancy information is not always easy to record reliably. In recent years, millions of people have been displaced by human and/or physical-induced disasters such as conflict in Syria or drought in the Horn of Africa. This makes accurate HDI data near-impossible to collect.
Gender inequality index (GII)	• The GII is a composite index devised by the United Nations. It measures gender inequalities related to three aspects of social and economic development. • Its ingredients are: reproductive health (measured by maternal mortality ratio and adolescent birth rates); empowerment (measured by parliamentary seats occupied by females and the proportion of adult females and males aged 25 years and older with some secondary education); labour force participation (rate of female and male populations, aged 15 years and older, in the workforce).	• Internationally, views differ on the validity of gender inequality as a development measure. Indeed, some states do not allow women to stand for election to parliament, including Kuwait. In Pakistan's Swat Valley, Taliban militia have burned down girls' schools. Cultures that do not support equal rights for women are unlikely to value the GII as a valid development measure. • Collecting reliable data on labour force participation rate may be tricky due to the numbers of women who work in the informal sector or under 'zero hours' contracts.

5.1 Development opportunities   45

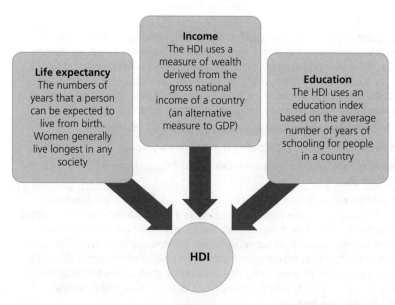

> **Keyword definition**
> **Informal sector** Unofficial forms of employment that are not easily made subject to government regulation or taxation. Sometimes called 'the black economy' or 'cash in hand' work, informal employment may be the only kind of work that poorly educated people can get.

Figure 5.5 Calculating the human development (HDI) index

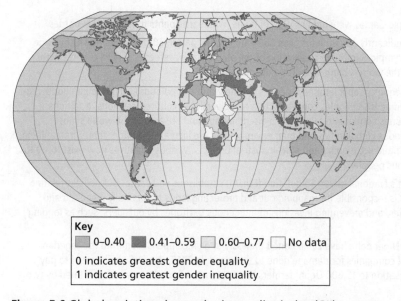

Figure 5.6 Global variations in gender inequality index (GII) scores 2014

■ Empowering women and indigenous or minority groups

In addition to uneven levels of human development that can be mapped at the global scale, disparities often exist between different groups of people *within* individual countries. Social disparities in wealth, access to education and political freedom persist in many states. Women, **indigenous people** and minority groups are disproportionately affected. Table 5.2 shows examples of affirmative action that aim to address these disparities and ensure that all individuals and communities within a society benefit equally from the human development process.

Table 5.2 Examples of affirmative action

> **Keyword definition**
> **Indigenous people** Ethnic groups who have enjoyed the uninterrupted occupation of a place for long periods of time (predating any arrival of more recent migrants).

Figure 5.7 Malala Yousafzai addressing the United Nations about education for girls

Affirmative action in support of:	
Women's right to an education	When she was shot by a Taliban gunman in 2012, Malala Yousafzai was actively campaigning for the right of girls to attend schools in the Swat Valley region of Pakistan (Figure 5.7). The campaign of violence against girls attending school meant that many have been denied an education, despite it being a fundamental right for the citizens of Pakistan. The attack on Malala shocked Pakistan and the world. She has since become a symbol of resistance against terrorism and the persistence of extreme social views that would deny women an equal right to education. Recently, efforts to improve the situation have taken place at varying scales: • Local schools in Pakistan's Federal Administered Tribal Areas (FATA) staged a day of action. • Pakistan's government has promised to improve participation of girls in primary school. • Pakistan is committed to the new United Nations Sustainable Development Goals (Figure 5.4), which include targets for education and gender equality.
Disabled people's participation in sports	Cultural attitudes towards disability are changing on a global scale. 1983–92 was the 'Decade of Disabled Persons'. The UN Convention on the Rights of Persons with Disabilities seeks to bring cultural change on a global scale in line with the Universal Declaration of Human Rights (UDHR). The UN has reaffirmed 'the universality, indivisibility, interdependence and interrelatedness of all human rights and fundamental freedoms and the need for persons with disabilities to be guaranteed their full enjoyment without discrimination'. It was not always the case that disabled people enjoyed equal rights, however. In the USA, sterilization programmes that sometimes targeted disabled people lasted until well into the twentieth century. Since then, a seismic shift in cultural attitudes has taken place in the USA and elsewhere. • International sporting events specifically for those with disabilities first began in 1948 with Second World War veterans participating. • The first official Paralympic Games were held in 1960 in Rome, with participants from just 23 countries. • The event has grown significantly and athletes from 164 nations took part in 2012. Global media coverage has helped turn the Paralympic Games – a celebration of the physical achievements of elite athletes with disabilities – into one of the world's biggest sporting events.
Indigenous people's access to land and resources	As many as 370 million indigenous people (around 5 per cent of the world's population) can be found worldwide. In more than 70 countries, indigenous peoples contribute to a rich diversity of cultures, religions, traditions and languages. Yet, these groups continue to face serious discrimination and are among the world's most marginalized peoples. Indigenous communities and their environments are increasingly under threat from mining, oil extraction, dam and road building, logging and agro-industrial projects. Various initiatives now exist to try to protect the rights of indigenous people: • In Brazil, the government's National Indian Foundation (FUNAI) establishes and carries out policies relating to indigenous peoples. It is responsible for mapping out and protecting lands traditionally inhabited and used by these communities, and preventing invasions of indigenous territories by outsiders, such as logging companies. • The Ogoni people of the Niger delta have campaigned tirelessly to gain compensation from the Nigerian government and large oil companies for damage done to their land: in 2015, the TNC Shell agreed to pay US\$70 million in compensation to 15,600 Ogoni farmers and fishermen whose lives were devastated by two large oil spills. • In recent decades, court cases in Canada have gradually helped to define and protect the land rights of indigenous First Nations people (see also Unit 5.2, page 71).

CASE STUDY

AFFIRMATIVE ACTION BY TNCS TO HELP CLOSE THE DEVELOPMENT GAP

Some of the world's largest TNCs have taken affirmative action to try to protect the human rights of LGBT (lesbian, gay, bisexual and transgender) communities in some countries. A lack of equality for LGBT people can be viewed as an example of a development gap which – along with lack of equality for women or minority ethnic groups – occurs sometimes *within* particular countries.

■ Many TNCs are headquartered in world cities such as San Francisco and New York. The academic Richard Florida has noted that global hubs such as these are often creative places where people are perhaps more open to new ideas and diversity. Some of these cities' technology and banking companies have led the way in trying to tackle prejudice in the workplace; a few have openly gay senior managers, including Apple's CEO Tim Cook (Figure 5.8).

■ Recently, 68 US companies – including Apple, Bloomberg, Cisco, Dupont, eBay, Gap, IBM, LinkedIn, Microsoft and Nike – appealed to the US state of North Carolina to abandon plans to stop transgender people from using bathrooms that did not match the gender of their birth.

However, when the US bank Goldman Sachs ran an LGBT recruiting and networking event at its Singapore office, a Singaporean government minister criticized the company for failing to respect local culture and context: 'They are entitled to decide and articulate human resource policies, but should not venture into public advocacy for causes that sow discord…' In Singapore – and many other countries in Asia, Africa and the Middle East – gay sex is still illegal.

For TNCs with a strong record of promoting diversity in the workplace in the countries they originate from, it can be a challenge to maintain a consistent global diversity policy. Campaigning too hard for equality could also affect a TNC's freedom to operate in some states; therefore affirmative action to tackle this development gap could become economically harmful. HSBC bank has made a statement about this issue: 'We respect the law in countries in which we operate, but that doesn't prevent us having a global point of view. And our global point of view is to be very strongly, very firmly on the side of diversity and inclusion.'

Sometimes, governments and civil society organizations have taken affirmative action in support of LGBT rights too.

- In 2013, Uganda's government announced a new law making homosexuality punishable by death or life imprisonment.
- At the time, the UN High Commissioner for Human Rights said: 'It is extraordinary to find legislation like this being proposed more than 60 years after the creation of the Universal Declaration of Human Rights'.
- Shortly after, Uganda lost millions of dollars of international aid when donor countries cancelled payments in protest.
- The Ugandan government has since scrapped the death penalty but homosexuality still warrants a prison sentence.

Figure 5.8 Tim Cook is Apple's Chief Executive Officer (CEO)

PPPPSS CONCEPTS

Think about how some cultural changes, such as embracing diversity, are contested. Are these 'Western' values or universal human values that all places ought to adopt?

Social entrepreneurship approaches to human development

Well-intentioned international aid has sometimes attracted criticism on the grounds that it does not solve poverty but perpetuates it instead. In Unit 4.2, for instance, we learned that charitable donations of clothes to Zambia had inadvertently devastated the country's own fledgling textiles industry. The unanticipated costs of charitable donations in particular contexts do not undermine the case for aid entirely, of course. International aid often plays an essential role in helping societies to recover following a disaster. Community resilience was pushed to its limits in the Philippines by Typhoon Haiyan in 2013; without international assistance, thousands more people might have died.

Attempts to foster long-term human development must do more than provide hungry or cold people with food and clothes, however. In order to eradicate poverty, strategies are required which can help people 1) develop capabilities and skills, and 2) find money to invest in technology or other resources needed to help them prosper economically. A broad mix of developmental strategies exists to help communities achieve these goals. Collectively, they are called **social entrepreneurship approaches** to human development. Examples include microfinance organizations and ethical trading initiatives.

> **Keyword definition**
>
> **Social entrepreneurship approaches** A way of trying to meet human development goals, which draws on business techniques and principles. Unlike large-scale 'top-down' lending to states, social entrepreneurship approaches often require only small (but potentially life-changing) loans to be made available to individuals and local communities.

■ Microfinance organizations and networks

Small loans of money to low-income borrowers are termed microloans: for example, a subsistence farmer who needs an injection of capital to expand their farm output to produce a tradable surplus. The theory behind this approach is that it will promote entrepreneurship, increase incomes in the long term and raise communities out of poverty effectively. Microloans are needed because subsistence farmers find it so hard to escape poverty. They lack capital because they can only grow enough food for their own needs. The seeds they use do not always yield good enough crops. The soil they plant them in may not be fertile. This is where microloans come in: Figure 5.9 shows how they help promote economic and social development.

Microloans provide farmers with the vital injection of cash their families need if they are to escape a cycle of poverty. A microloan is not a 'free hand-out' and must be paid back, however. One advantage of a small commercial loan like this, when compared with charitable aid, is that poor people feel they can stand on their own two feet instead of being dependent on others.

- The best-known provider of microloans is the Grameen Bank in Bangladesh. It has lent money to 9 million people, 97 per cent of whom are women.
- This gendered pattern of lending dates back to the bank's early days in the 1980s, when it found that women had higher repayment rates and tended to accept smaller and less riskier loans.
- Subsequently, many microcredit institutions have used the goal of empowering women to justify their disproportionate loans to women.
- In contrast to the billions of dollars lent to countries, microloans involve just a few hundred dollars. But they can play a crucial role in jump-starting development at a local level. The theory is that if enough villages are helped then, in time, an entire country can develop.

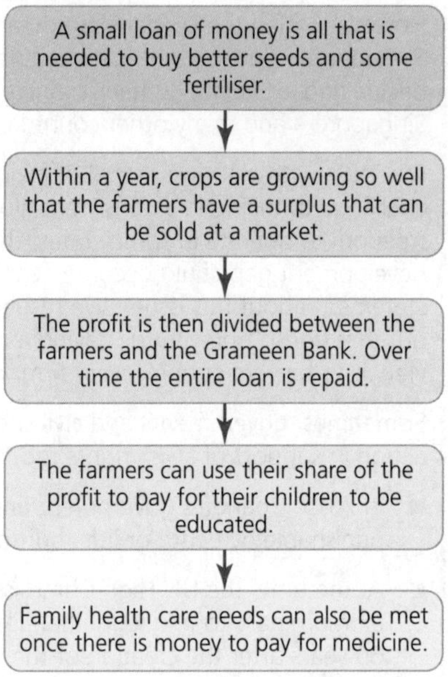

```
A small loan of money is all that is
needed to buy better seeds and some
             fertiliser.
                │
                ▼
Within a year, crops are growing so well
that the farmers have a surplus that can
         be sold at a market.
                │
                ▼
The profit is then divided between the
farmers and the Grameen Bank. Over
   time the entire loan is repaid.
                │
                ▼
The farmers can use their share of the
profit to pay for their children to be
              educated.
                │
                ▼
Family health care needs can also be met
once there is money to pay for medicine.
```

Figure 5.9 Microloans flowchart

■ Microlending in Malawi

Microlending has transformed the lives of subsistence farmers near the shores of Lake Malawi, in the Republic of Malawi, a land-locked tropical country in southeast Africa. The farmers have also become more deeply integrated into global networks as a result of the lending. Malawi is among the world's least developed countries, with a low life expectancy and a high infant mortality rate. Recently, microloans have been used to help Malawian farmers buy new seed types that yield more crops, and fertilizer too. The loans are small enough that people can safely repay them in instalments, provided the money has been wisely invested. The source of the lending is a mega-farm project run by an entrepreneur called Duncan Parker.

- The mega-farm offers subsistence farmers microloans as part of a wider business opportunity.
- By growing improved red pepper crops, the farmers have food for themselves and can sell a large surplus to the mega-farm.
- A total of 8,000 small farms now use the mega-farm as their distribution hub.
- The mega-farm processes the red peppers the farmers sell to make paprika, which is sold to the TNC Nando's restaurants in Europe.
- Growing numbers of the farmers' children attend school now.

■ Does microlending always work?

Microloans are not an unqualified success story, however.

- Some studies show mixed results and there is evidence of lending sometimes being more effective in helping existing businesses to prosper rather than helping entirely new ones to thrive.
- Loans have been used in some cases to purchase farm technology that reduces the need for farm labour (and so creates unemployment, thereby driving other people back into poverty).
- There are cases also of loans being misused: families may use a microloan to cover burial costs for a relative, or to pay larger dowries for their daughters, and are unable to pay the money back later.
- Some loans to women have been misappropriated by husbands or male relatives. One study in Bangladesh reported that this problem became more common when bigger loans were involved.

> **PPPPSS CONCEPTS**
>
> Think about the spatial interactions that feature in this example of microlending. A network of subsistence farmers in Malawi have become part of the global supply chain network for a leading TNC restaurant chain.

Alternative trading networks

You will probably be familiar already with the general principles of 'fair trade' (or 'alternative trade'). The work of the Fairtrade Foundation in particular has been very important for some communities in developing countries. The aim is to give producers a fair fixed price for the goods they produce. If the global price for a particular crop like coffee collapses, Fairtrade farmers will continue to receive a steady income which safeguards their quality of life. It is important to recognize that this remains a profitable business model however, and not a loss-making or charitable endeavour.

The venture succeeds because a sufficient number of shoppers in high-income countries have increased what they spend on Fairtrade food and goods over time. Examples of Fairtrade produce include chocolate, bananas, wine and even higher-value clothing items such as jeans and footballs. Shoppers are motivated by 1) a sense of ethics and 2) genuine curiosity about the provenance (origin) of the goods they buy. Some consumers are happy to pay a little more, knowing that a higher proportion than usual of the bill will find its way directly into the pay packets of some of the world's poorest people.

The benefits of the Fairtrade system can be shown using the example of a village called Chagelen in the north-east of Punjab province in Pakistan.

- In its promotional literature, Fairtrade explains how 18-year-old Sameena Nyaz, like many other people in Chagelan, belongs to a Fairtrade collective.
- Sameena stiches Fairtrade footballs for a living. Although she did not attend school, her younger sisters are able to because of the enhanced family income.
- Chagelan villagers belong to a health-care scheme that is paid for by the Fairtrade system; social development goals are being met in Chagelen alongside higher earnings and wealth creation.

There are drawbacks to Fairtrade and the scale it can operate at, however.

- As the number of schemes grows, it becomes harder to ensure that money and benefits have been correctly distributed to all growers and manufacturers.
- It is simply not possible for all the world's farmers to join a scheme offering a high fixed price for potentially unlimited crop yields. If a guaranteed price was offered to anyone who wanted to grow an unlimited number of bananas, the entire system would collapse.
- The higher price of Fairtrade products means that many shoppers avoid buying them, especially during times of economic hardship. This also puts limits on the number of producer collectives that can become part of the scheme. Fairtrade sales continue to increase worldwide (Table 5.3) but the rate of growth has slowed since the Global Financial Crisis (GFC), reflecting the greater financial hardship many consumers in emerging and developed economies now face (see Unit 6.1, page 82).

Table 5.3 Fairtrade worldwide revenues 2004–14

Year	Revenue (in million Euros)
2004	832
2005	1,132
2006	1,623
2007	2,381
2008	2,895
2009	3,443
2010	4,319
2011	4,984
2012	4,787
2013	5,500
2014	5,900

Alternative gold networks

A vivid example of how alternative trading networks can make a difference is the international gold trade. Many of the tens of millions of people working in the artisanal and small-scale mining (ASM) informal sector risk disease, serious injury and death. ASM miners are often taken advantage of by unscrupulous middlemen, according to the Fairtrade Foundation and the Alliance for Responsible Mining (ARM). There are three interrelated concerns:

- *Pay, health and safety* There are over six times the number of accidents in ASM compared with large-scale mining, mainly due to its larger labour force – which may include young children – and poorer working conditions.
- *Environmental issues* Many examples can be cited of gold mining's negative impacts including deforestation and land degradation through air, water and soil pollution from toxic chemicals used to process gold ore, including mercury, cyanide and nitric acid. Around 80 per cent of all human mercury poisoning is caused by artisanal gold mining.

- *Conflict* Poverty pushes many people into working in artisanal mines. But some are actively forced to do so by militias operating in conflict zones, notably in Democratic Republic of the Congo.

Fairtrade and Fairmined gold have been available for several years. The Cotapata Mining Co-operative in Bolivia was the first Fairtrade and Fairmined conventional mining organization to be certified in 2011. Others have since been established in Colombia, Peru and Mongolia.

- To gain certification, artisanal miners in a region must first band together to form an organization that Fairtrade can deal with directly.
- Each organization pledges to participate in the social development of their communities by eliminating child labour (under 15) from their locality, introducing protective gear for all miners and recognizing the right of all workers to establish and join trade unions.
- They must also use safe and responsible practices for management of toxic chemicals in gold recovery, such as mercury and cyanide.

Alongside Fairtrade, major global jewellery companies have established alternative trading networks which exclude suppliers who do not meet their social and ecological standards. The famous jeweller Tiffany announced it would not buy gold from the controversial Pebble Mine in Alaska, for instance, because of the ecological impacts of mining operations there (Figure 5.10).

Figure 5.10 Civil society opposition to the Alaska's Pebble Mine led to Tiffany's decision to find alternative ethically sourced supplies of gold

■ TNC corporate social responsibility frameworks and global agreements

Large businesses increasingly accept the need for **corporate social responsibility**. As Unit 4.2 explains, the largest TNCs have thousands of suppliers to whom they outsource work; this increases the risk of 1) the violation of workers' human rights and 2) brand products becoming linked with workforce exploitation.

The TNCs used as case studies and examples earlier in this book – such as Fender, Dyson and Apple – have strict codes of practice which prohibit worker exploitation in 1) the offshored facilities they actually own and 2) their first tier of outsourcing suppliers. A code of conduct guarantees certain rights for employees and may cover legal areas such as:

- maximum number of hours an employee is required to work each week
- the right to belong to a trade union or other employee organization
- employee entitlement to benefits such as holidays or sick pay
- any right to compensation if an employee is sacked before a contract ends.

> **Keyword definition**
>
> **Corporate social responsibility** Recognizing that companies should behave in moral and ethical ways towards people and places as part of their business model.

However, Apple learned in 2011 that workers for its iPhone touchscreen supplier, Lianjian technology, had been poisoned by a chemical cleaning agent. Apple was unaware of the working conditions at Lianjian because it is a third-tier supplier (Figure 4.24, page 23). Large TNCs do not generally monitor the lower tiers of their own supply chains; indeed, it may be almost impossible for TNCs to monitor the work and pay conditions for the workforce of every single outsourcing supplier that is part of their extensive global production network (Figure 5.11).

- In 2006, UK supermarket Tesco banned the use of cotton from Uzbekistan in all of its items of clothing, following reports of child labour.
- However, Tesco has an outsourcing relationship with textile companies in Bangladesh, Turkey and China and it took them many more years to finally eliminate Uzbek cotton from the supply chains supporting these companies.

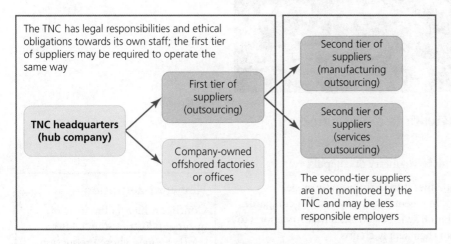

Figure 5.11 The moral and ethical geography of a TNC's global production network

■ A global agreement for Bangladesh

The Accord on Fire and Building Safety in Bangladesh is a significant recent development that shows TNCs taking greater responsibility for working conditions in their supply chains. It was introduced following the collapse of the Rana Plaza building in Dhaka, Bangladesh in 2013 (Figure 5.12).

- This disaster led to the deaths of 1,100 textile workers.
- On the day of the collapse, workers were sent back into the building by Rana Plaza's managers to complete international orders in time for delivery, even though major cracks had appeared overnight in the building.
- Wal-Mart, Matalan and other major TNCs regularly outsourced clothing orders to Rana Plaza.

Since then, many European TNCs have signed the Accord, which is a legally binding agreement on worker safety. These companies now promise to ensure safety checks are carried out regularly in all Bangladeshi factories that supply them with clothes.

- The Accord states that signatory TNCs are 'committed to the goal of a safe and sustainable Bangladeshi Ready-Made Garment (RMG) industry in which no worker needs to fear fires, building collapses, or other accidents that could be prevented with reasonable health and safety measures'.
- The agreement covers all suppliers producing products for the TNCs. It requires these suppliers to accept inspections and implement remediation measures in their factories if unacceptable safety risks are discovered.

Many TNCs that outsource work to Bangladesh have not signed up yet, however, despite the fact that the Accord will result in as little as 2 cents being added to the production cost of a T-shirt. Profit maximization dominates the decision-making of the non-signatory companies. In contrast, the H&M clothing company not only signed the Accord but has additionally campaigned for Bangladesh's government to strengthen safety-at-work legislation and to increase the country's minimum wage.

Figure 5.12 The collapse of the Rana Plaza clothing factory in Bangladesh in 2013; Rana Plaza had an outsourcing relationship with many major TNCs

■ A social responsibility framework to help Democratic Republic of the Congo (DRC)

Corporate social responsibility actions may be voluntary or compulsory.

- TNCs can introduce new ethical guidelines into their business model voluntarily because it is right to do so. In recent years, a handful of companies such as Intel, HP and Apple have voluntarily led electronics industry efforts to try to eradicate **conflict minerals** from their supply chains.
- Alternatively, TNCs may be compelled by law to comply with social responsibility legislation in the states in which they are domiciled or operate.

> **Keyword definition**
> **Conflict minerals** Products of mining industries sourced from conflict zones whose production may have involved slave labour.

The US government's Dodd–Frank Wall Street Reform Act of 2010 is an example of forced compliance: it is a 'top-down' approach to corporate social responsibility. Dodd-Frank is a law which made it illegal for US-registered TNCs to make use of 'conflict minerals' thought to have originated in Democratic Republic of the Congo (DRC). This is a country where millions of lives have been lost in a conflict partly fuelled by competing ownership claims over natural resources (the issues affecting DRC were explored in Unit 4.3, page 37). Dodd–Frank requires US companies to find out where their 3T (tin, tungsten and tantalum) and gold minerals are from and then disclose whether or not those minerals funded armed groups.

Most end-user companies did not know their sources of minerals before Dodd–Frank, so the law has forced them to look deeper. One view is that this has therefore had a positive effect. Militia groups now find it harder to sell gold and diamonds to fund their wars. However, there is an opposing view that Dodd–Frank may have inadvertently worsened poverty and instability in parts of DRC:

- This is because some TNCs have responded by avoiding exports from DRC altogether, including legitimate supplies, especially when minerals are readily available from other countries. Companies want to avoid any risk whatsoever with becoming associated with conflict minerals.
- As a result, a well-intentioned 'solution' is now part of the enduring problems of poverty and poor global connectivity for DRC and its people.
- For some small mining companies and cooperatives, Dodd–Frank has been disastrous. In South Kivu Province, many small-scale miners can no longer find buyers for the conflict-free metal ores they have dug out, despite living in what is now a conflict-free region.

5.1 Development opportunities 53

■ **KNOWLEDGE CHECKLIST**

- The concept of human development
- The UN Sustainable Development Goals criteria
- Ways of measuring human development, including the human development index (HDI) and gender inequality index (GII)
- The validity and reliability of different development measures
- The importance of the empowerment of women and indigenous or minority groups, using detailed examples of affirmative action to close the development gap (TNC action in support of LGBT rights)
- The importance of social entrepreneurship approaches for human development
- The work of microfinance organizations
- The importance of alternative trading networks (Fairtrade)
- TNC corporate social responsibility frameworks
- Global agreements on social responsibility (Bangladesh Accord on Fire and Building Safety)

EVALUATION, SYNTHESIS AND SKILLS (ESK) SUMMARY

- How views may differ on which aspects of human development and rights are most important
- How human development is supported by the actions of stakeholders at different geographic scales

EXAM FOCUS

ANALYTICAL WRITING SKILLS

Below is another sample HL answer to a part (a) exam-style extended writing question, which asks you to apply your knowledge and understanding of a fairly narrow topic (under assessment criteria AO1 and AO2). The mark scheme is shown on page vi.

This time, the command word is 'explain' instead of 'analyse'. Read it and the comments around it.

Explain why it might be hard to establish how human development varies between different countries. (12 marks)

Human development can be defined as the range of ways in which societies change as they gain more wealth and so experience improved health, education and often political rights too. In the world today, countries differ greatly in their level of development when measured in terms of average income per person and life expectancy. However, can we really trust the data? There are many reasons why human development measurements may not be accurate. (1) Firstly, take the Human Development Index (HDI). HDI scores are an average of a country's global ranking for three categories. These are gross domestic product per capita, life expectancy and literacy rates. (2) In poor countries it can be hard to collect accurate data for all of these. For example, in Africa people have not got proper jobs and work in the informal sector which is not recorded properly by the government. (3) Therefore it is hard to know how much people really earn. In many poorer countries millions of people move to cities every week. This makes it almost impossible for a government to collect accurate data about its population because people are constantly on the move. Also information can go out of date quickly, particularly in a country which is developing rapidly such as China. (4) Another important way of measuring human development is the gender inequality index (GII). This shows how fairly women are treated in different countries and what rights they have. In some countries women cannot vote (Kuwait) and in others it can be difficult for girls to get an education. Malala Yousafzai was shot near her home in rural Pakistan because she promoted education for girls using a website that the BBC set up for her. Many people in that part of the world still think that girls are second-class citizens and this could affect Pakistan's GII score. (5)

Therefore there are many reasons why it is hard to measure how human development varies between countries.

Examiner's comment

This piece of extended writing lacks structure (AO4), which makes it harder to read and mark. The writer applies (AO2) a few relevant ideas but with very few good examples or detail. The knowledge shown (AO1) is only partial because no distinction is made between validity and reliability. Overall, this would only reach around half marks.

1. This is a sound introduction, which defines the key idea of human development. However, there is no attempt to explain what 'hard to establish' could mean (a contrast could be offered between validity and reliability issues for instance).

2. This is not a very accurate account of HDI, despite it being a key content area of the subject guide. No examples are given of actual HDI scores.

3. While it is good to see knowledge of the informal sector, the remark that 'in Africa people have not got proper jobs' is poor. Can you see why?

4. This is a good point and it is an appropriate example.

5. Could you see how the writer is failing to answer the question? This paragraph is a missed opportunity to explain
1) why it might be hard to collect reliable inequality data and
2) why some people still do not support gender inequality and so would argue this is not a valid measure of development.

Avoiding generalizations

Very few things will annoy an examiner more than the comment: 'For example, in Africa'. This is a continent composed of more than 50 states and many hundreds of ethnic groups and nations. Examples should always refer to specific countries and cities and include supporting data wherever possible.

5.2 Changing identities and cultures

Revised ☐

The word 'culture' describes what writer Raymond Williams called a society's 'structure of feeling'. Various shared **cultural traits** are held in common by different local or national societies (Figure 5.13). Cultures change and evolve over time naturally; global interactions have accelerated the rate of cultural change for many places, however. In addition to exploring how globalization influences culture at varying scales, this unit also examines how perspectives differ on both the degree of change that is occurring, and its desirability.

> **Keyword definition**
> **Cultural traits** Culture can be broken down into individual component parts, such as the clothing people wear or their language. Each component is called a 'cultural trait'.

Language: Some countries have a single national language with local dialects, or several languages belonging to different indigenous ethnic groups

Food: National dishes and diet traditionally reflect the crops, herbs and animal species that are available locally

Clothing: National and local traditions may reflect traditional adaptations to the climate (such as wearing fur in polar climates) or religious teachings

Religion: There are several main world religions, each with its own local variants; religion is an important cultural trait that also informs food and clothing, and may be highly resistant to change

Traditions: Everyday behaviour and 'manners' are transmitted from generation to generation, from parents to their children, such as saying 'thank you' or shaking hands

Cultural traits

Figure 5.13 Cultural traits

The global spectrum of culture

Revised ☐

Globally, a wide spectrum of cultural traits, **ethnicity** and **identity** has always existed. Global interactions mean that **cultural diversity** is changing, however. We can identify aspects of a shared global culture that is spreading perhaps at the expense of local indigenous cultures in certain places. In addition, we may observe how local cultures everywhere show signs of change over time, often in response to powerful global forces. A useful way to begin an analysis of the global spectrum of culture is to ask: why have certain cultures – notably that of the USA – gained such influence on a global scale? How is this power achieved?

■ Cultural diversity at the global scale

Is there such a thing as a **global culture**? Although individuals and societies differ from one another in a great many ways, there are also plenty of things that a large proportion of the world's population have in common. A typical 'global citizen' could be someone under the age of 30 who wears jeans, listens to rock or rap music, uses social media on an iPhone or similar device, enjoys buying branded clothing from the local shopping mall and can – when required to – hold a conversation in French, English, Spanish, Arabic or Mandarin. You can reflect on the extent to which people where you live participate in some or all of these behaviours.

Over time, the cultural commonality of different places has often increased as a result of the global flows and networks discussed in Units 4.2 and 4.3. Table 5.4 shows how the factors and processes responsible for heightened global interactions have also helped to shape a so-called global culture. You will notice that the sources of global cultural influences are shown to include not only North American and European countries but also China, India, Russia, Qatar and Japan, among others. This is because it is increasingly an over-simplification to view cultural globalization, **Westernization** and **Americanization** as being essentially interchangeable ideas.

- Western cultures continue to have a disproportionate global influence: the wide use of European languages around the world is a reflection of the enduring wealth and power of these countries and their TNCs.

> **Keyword definitions**
> **Ethnicity** The shared identity of an ethnic group, which may be based on common ancestral roots or cultural characteristics such as language, religion, diet or clothing.
>
> **Identity** An individual or society's sense of attachment to one or more places. This may be at the country, region, city or village scale. People may feel a sense of belonging to multiple places due to their family history or because of the differing loyalties or attachments that can operate at the state level (for example, a sense of patriotism or nationalism) and local level (for example, family 'roots' in a particular neighbourhood or support for a local sports team).
>
> **Cultural diversity** The level of heterogeneity (difference) exhibited by a community in terms of ethnicity, religion, language or other defining cultural traits. Cultures that lack diversity instead show homogeneity (sameness).

- It is arguably the case that the USA has greater cultural influence than any other state: this is an important part of its soft power (see Unit 4.1, page 5).
- However, the rise of some newly powerful states in Asia, the Middle East, South America and Africa has meant that there are plenty of 'non-Western' ingredients to what we call global culture, as Table 5.4 shows. The diffusion of culture at a global scale owes much to TNCs; but as the number of non-Western global brands continues to rise, it is unsurprising to see an increasingly wide range of ingredients being added to the global cultural mix.

> **Keyword definition**
> **Global culture** A shared sense of belonging at the planetary scale that is demonstrated through common ways of communicating, consuming media and food, dressing or behaving (including shared social norms such as a commitment to upholding human rights).

Table 5.4 Ways in which powerful countries and companies spread their culture globally

Factor	TNCs	Global media	Migration and tourism
Influence	The global dispersal of food, clothes and other goods by TNCs has played a major role in shaping a common culture. Some Western corporations, like Nike, Apple and Lego, have 'rolled out' uniform products globally, bringing cultural change to places. Asian technology companies such as Samsung and Huawei have an increasingly important global influence on patterns of entertainment and social networking.	Media giant Disney has exported its stories of superheroes and princesses everywhere, along with stories of Christmas (originally a Western Christian festival). The BBC continues to help the UK maintain its high level of global cultural influence. Important non-Western influences on global culture and media include India's Bollywood and Qatar's Al Jazeera TV channel. Japanese children's TV is influential, notably Pokémon.	Migration brings great cultural changes to places. Europeans travelled widely around the world during the age of empires, taking their languages and customs with them. Today, tourists introduce cultural change to the distant places they visit. Western tourists have helped diffuse Western cultural values to different places; increasingly, Chinese and Indian tourists are spreading their own culture.

One particularly striking manifestation of a 'global culture' is the continued rise of a few languages at the expense of many less well-known ones. Globalization and global urbanization have meant that a few global languages increasingly dominate:

- 95 per cent of the world's population speaks one of just 400 languages, each spoken by millions of people
- 40 per cent of people speak one of just eight major languages: Mandarin, Spanish, English, Hindi, Portuguese, Bengali, Russian and Japanese.

At the same time, one in four of the world's 7,000 minor languages are now threatened with extinction (Figure 5.14); half now have fewer than 10,000 speakers, and these 3,500 languages are spoken by only 0.1 per cent of the world's population (8 million people).

- Linguistic diversity is declining worldwide as fast as biodiversity – about 30 per cent since 1970.
- Papua New Guinea originally had around 1,000 indigenous languages; but as global interactions accelerate, the physical, technological and economic barriers that once allowed so many isolated languages to develop are being removed. Deforestation is the greatest threat: it has led to the forced migration of entire forest communities.

> **Keyword definitions**
> **Westernization** The imposition and adoption of a combination of European and North American cultural traits and values at a global scale.
>
> **Americanization** The imposition and adoption of US cultural traits and values at a global scale.

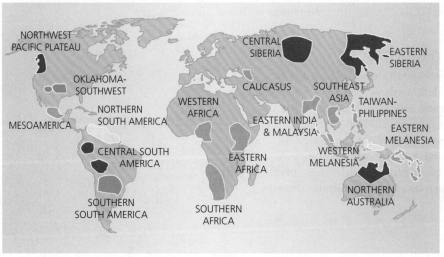

Figure 5.14 The distribution of threatened and disappearing languages

Existing alongside the eight major languages is a widely spoken subset of English called 'Globish'. In 1995, Jean-Paul Nerrière first used the term to describe a stripped-down vocabulary consisting of just 1,500 English words but spoken by up to 4 billion people. The Globish 'micro-language' is distinct from more complex variants of true English spoken as an official language in the USA, Australia, Canada and elsewhere. It serves a purely utilitarian purpose by providing global citizens with the means they need to exchange vital information with one another, such as travel directions or terms of business. Global citizens are people who are routinely involved in global interactions, including:

- tourists, international migrants and the international business community
- residents in global hubs (megacities such as Los Angeles or São Paulo), where many different ethnic and migrant groups require a common language for communication
- social network users, such as Facebook members, who correspond online with people of many nationalities.

Globish has a long history of being adopted by 1) international migrants arriving in English-speaking countries such as the USA and 2) the citizens of more than 60 ex-British colonies. Since the 1990s, however, Globish has diffused into countries that traditionally lack a strong affinity with British or American culture, such as Japan, China or Brazil. This is because English:

- has dominated internet communication from its outset
- has enjoyed a supra-national rise as the global language of business (commerce, technology and education), in part due to the English-speaking USA's superpower status.

It is, however, important to think critically about whether the spread of Globish is linked causally with the decline of other languages. With its highly limited vocabulary, Globish is not *replacing* other languages; instead, people adopt it *in addition to* their native tongues. We should, therefore, be cautious about viewing Globish as a direct cause of reduced global cultural diversity.

Also, Globish exists in many different local variant forms and arguably even contributes to linguistic diversity. Words, syntax and grammar vary from country to country because of the way English has blended with different native languages – 'Singlish' is the Singaporean variant of Globish, for instance, while 'Hinglish' is the version used by Hindi speakers. Although these localized versions of Globish differ, there is enough common ground for people of different national origins to stage a functional conversation when they meet.

It is worth noting also how Globish may contribute to the enduring soft power of both the USA and UK (see Unit 4.1, page 5). British music acts and Hollywood films enjoy large global audiences in part thanks to their easy decipherability for Globish speakers. The governments of both the UK and USA openly acknowledge that their culture industries are a key means for maintaining both nations' soft power on the global stage.

Cultural diversity at the national scale

Some states have a strong single cultural identity while others do not. Often states are home to a range of different ethnic, religious or national communities (insofar as some states house numerous different cultural 'nations' or homelands). The degree of national cultural complexity varies greatly from state to state. Unit 4.3 (page 38) explores the relative isolation of some island communities such as Iceland, and the cultural homogeneity this may support. In contrast, Singapore's geographical location has combined with liberal migration rules to encourage migration and the cultural diversity it brings.

The cultural mix of the USA is extremely complex. Prior to the arrival of Europeans, the continent of North America was home to a heterogeneous mix of indigenous peoples, including the Sioux and Nahavo tribes. Today, 320 million people occupy the same territory; the majority are the descendants of a global mix that includes Italians, Greeks, Scandinavians, Scots, Irish, Mexicans, Cubans, Indians, Pakistanis, Vietnamese, Puerto Ricans, Koreans and many more besides. Yet from this mix a so-called 'American culture' has developed over time, in part due to a **melting pot** effect (Figure 5.15). American culture is both inclusive (allowing new arrivals to participate) and dynamic (it becomes modified in turn by each wave of new arrivals).

> **Keyword definition**
>
> **Melting pot** A cultural process that involves different communities mixing over time to form a more uniform culture which combines traits drawn from the traditions of each of the original communities.

Figure 5.15 US school children share a political identity but come from families with diverse cultural backgrounds

Some of the world's more recently established ex-colonial states are also home to a broad mix of cultures. However, not enough time has passed yet for them to develop a shared cultural identity as a national people.

Prior to today's 'shrinking world' era, mountain ranges and rivers sometimes formed natural barriers to population movements and provided the basic geometry for nations to develop in particular places. Over time, long-settled ethnic groups formed a strong association with their land. In Europe, borders formed organically over centuries or millennia, for instance. Today's European geopolitical map corresponds broadly with its cultural and linguistic map.

In Africa, the situation is very different, however. Political boundaries correspond poorly with the distribution of different cultural and ethnic groups. This is a legacy of the partition of Africa in haste by European nations in the eighteenth and nineteenth centuries. Space was divided between competing powers. For instance, the boundary between Egypt and Sudan is a straight line drawn by Great Britain in 1899. It is part of the 22nd parallel north circle of latitude. Like other boundaries drawn by Europeans, it paid little or no consideration to the people actually living there. The colonial powers were more concerned with dividing up Africa's raw materials and water resources among themselves. It is hard to imagine a worse approach to culturally cohesive state building.

- By 1900, many African ethnic groups found themselves living in newly formed nations that in no way represented their own heritage.
- Some long-established ethnic regions were split or divided into two or more parts, with each becoming part of a different newly established territory.
- Figure 5.16 shows how culturally diverse most states are as a result of European map-making and the dividing of ethnic 'homelands'. Nigeria alone is home to more than 250 different ethnic groups and would be culturally complex irrespective of more recent cultural exchanges associated with contemporary global interactions.

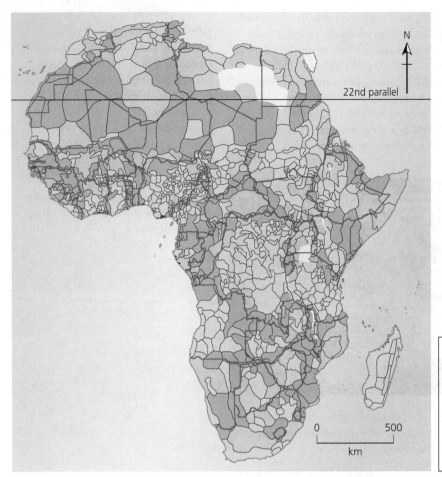

Figure 5.16 An ethno-linguistic map of Africa with modern nation states superimposed to show their inherent cultural diversity

Cultural diversity at the city scale

Migration flows both in the past and current era have brought great changes to the cultural and ethnic composition of world cities. Figure 5.17 shows prime London neighbourhoods where significant amounts of property have passed into the ownership of a particular group of foreign nationals. Competition from rich overseas 'asset' buyers has been a major contributing factor to the stratospheric rise in house prices in London and many other world cities since 2000. A record sale price of around US$200 million was achieved for a flat in London's fashionable One Hyde Park development in 2014: the anonymous buyer was widely believed to be Russian.

As a result of French and Russian acquisitions of property in central London, neighbourhood character is changing to create new **ethnoscapes**. Restaurants have begun to serve pickled herrings, snails and other decidedly 'non-British' menu items. The number of French nationals now living in London makes it 'France's sixth biggest city' in terms of population: the French government believes there are between 300,000 and 400,000 citizens living in London, exceeding the inhabitants of Bordeaux, Nantes or Strasbourg. Elsewhere in London are communities drawn from almost every country in the world, from Somalia to Romania. Communities with a particular national background may be further subdivided according to religion: Indian Muslim and Sikh communities have their own residential patterns in London.

Two important points should be noted when thinking critically about this topic:

1 Although global interactions are associated with the lessening of cultural diversity at a global scale, we are also seeing unprecedented levels of cultural diversity *within* particular world cities that are hubs for international migration. The answer to the question 'What is the effect of global interactions on cultural diversity?' therefore varies *according to what geographical scale we are looking at*.

Keyword definitions

Ethnoscape A cultural landscape constructed by a minority ethnic group, such as a migrant population. Their culture is clearly reflected in the way they have remade the place where they live.

Cultural landscape The distinctive character of a geographical place or region that has been shaped over time by a combination of physical and human processes.

2   History shows that levels of diversity at the local level may lessen *as time passes*. For instance, around 70 per cent of people in London identify themselves as being 'white British'. Yet this community was originally far from culturally homogenous. At different times in the past, varied white ethnic communities of Viking, Anglo-Saxon, Celtic, Roman and Norman descent have all lived in London. Over time, these diverse migrant groups combined in a cultural melting pot that gave rise to modern English culture and language as it is spoken today. It may be the case that in the future, cultural diversity will begin to lessen in world cities like London, Toronto and Paris that are currently incredibly diverse. This will be on account of cultural intermixing and intermarriage between the different migrant communities.

PPPPSS CONCEPTS

Think about the importance of timescale over which cultural processes operate. A very diverse place today could become homogenous in the future. Different cultural groups may mix over time to create a new uniform culture.

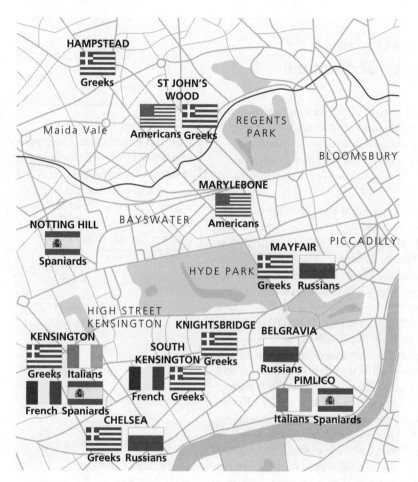

Figure 5.17 Some London neighbourhoods have a strong non-British identity: but will this lessen over time due to the melting pot effect?

The cultural effects of global interactions

Revised

What, if anything, can be done to protect the diversity of local cultures and languages from the spread of global culture? This theme is returned to in later units which explore migration policies and anti-globalization movements in varying national contexts. A separate question which can be addressed briefly here is: *should* anything be done? Some people view the cultural effects of global interactions – and the spread of so-called global culture – positively. This is an optimistic perspective on **hyperglobalization**, which is grounded in the belief that narrow and localized cultural identities have often given rise to chauvinistic nationalistic attitudes which may 1) fuel conflict and 2) jeopardize effective global collaboration to tackle pressing issues such as climate change.

Optimistic hyperglobalizers envisage a 'global village' where individual group attachments to ethnic and religious identity will be replaced by a shared identity based on the principles of global citizenship. In theory, this should increase global and local prospects for peaceful coexistence by reducing possible opportunities for prejudice to arise. Not everyone agrees with this view, however (Figure 5.18). The next part of this unit examines critically several cultural effects of global interactions that some individuals and societies celebrate but others resist.

Keyword definition

Hyperglobalization The theory of hyperglobalization proposes that the relevance and power of countries will reduce over time. Global flows of commodities and ideas may result ultimately in a shrinking and borderless world. There are, however, competing perspectives on the desirability of this.

Optimistic hyperglobalizers may view the loss of cultural diversity as being positive in certain ways. They believe in a modern culture of global citizenship that values equality and frees people from conflict or discrimination on the grounds of race, gender or sexuality. Some cultures still practise female genital mutilation, for instance. Optimistic hyperglobalizers see progress in the loss of localized cultural traditions like this.

Pessimistic hyperglobalizers may view the loss of local culture negatively. They are concerned that half of the world's languages are projected to disappear. With the death in 2010 of the last speaker of Bo, an ancient language in the Andaman Islands, India lost an irreplaceable part of its heritage. Pessimistic hyperglobalizers view this as an irreversible loss, like the extinction of a biological species. They argue that global cultural diversity must be valued and protected.

Figure 5.18 Diverging perspectives on hyperglobalization

■ The effects of cultural diffusion and cultural imperialism on places

Powerful civilizations have brought cultural change to other places for thousands of years. This spread is called cultural diffusion. Sometimes it is achieved through coercion, using legal or even military tools. Forced assimilation of culture is also called **cultural imperialism**. In the past, languages, religions and customs were spread around the world using force by the invading armies of the Persians, Romans and Portuguese, for instance.

> **Keyword definition**
>
> **Cultural imperialism** The practice of promoting the culture/language of one nation in another. It is usually the case that the former is a large, economically or militarily powerful nation and the latter is a smaller, less affluent one.

- Between approximately 1500 and 1900, leading European powers built global empires. The invasion and colonization of South America by Portugal and Spain was well underway by the 1600s. In the 1700s and 1800s, the pattern was repeated in Asia and Africa by Great Britain, France, Belgium and Holland. Traditional territorial names such as 'Kongo' and 'Akan' were overwritten on the map of Africa. Names like 'Belgian Congo' and 'Gold Coast' appeared in their place. In a remarkable act of hubris, modern-day Zimbabwe gained the name Rhodesia in the late 1800s in honour of a British man called Cecil Rhodes.
- As more and more territories came under their direct rule or influence, rival European states each built a global empire. Greatest in extent was the British Empire. By 1880, it held dominion over one-third of the world's land surface and one-quarter of its people. Empires were a vehicle for the diffusion of European languages, religions, laws, customs, arts and sports on a global scale. The British Empire was founded on exploration and communications technology including sea power and the telegraph (Figure 5.19). After independence, many ex-British colonies chose voluntarily to remain in the British Commonwealth. This grouping of nations retains a surprising sense of unity. To this day, several states and places round the world feature the Union Jack as part of their own flags, including Australia, Fiji, the Falkland Islands, Bermuda and Ontario (Canada).

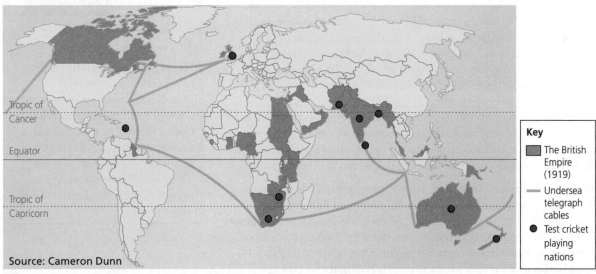

Figure 5.19 The network of undersea telegraph cables connecting the British Empire in 1910: the internet of the Edwardian era

While the age of empires may have ended, countries like the USA and UK continue to bring cultural change to other places through their use of soft power. No overt force is involved. Instead, these powerful, wealthy states shape global culture through their disproportionately large influence over global media and entertainment. Table 5.5 shows the 2015 Global Brand Index produced by WPP, a market research company. This can be analysed for evidence of the continued dominance of Western (European and North American) nations via their brands. There are also signs that non-Western brands are beginning to influence global culture too. Can you spot these?

Table 5.5 Top 30 Global Brands in 2015

1	Apple	11	Tencent	21	Baidu
2	Google	12	Facebook	22	ICBC
3	Microsoft	13	Alibaba	23	Vodafone
4	IBM	14	Amazon	24	SAP
5	Visa	15	China Mobile	25	American Express
6	AT&T	16	Wells Fargo	26	Walmart
7	Verizon	17	GE	27	Deutsche Telekom
8	Coca-Cola	18	UPS	28	Nike
9	McDonald's	19	Disney	29	Starbucks
10	Marlboro	20	Mastercard	30	Toyota

Some critics have argued that these global brands are a vehicle for a new neo-colonial phase of Western cultural imperialism; they say Western culture continues to be propagated globally via a global media over which the USA in particular has disproportionate influence. Take the Walt Disney Company, for instance: across Asia, young children encounter Mickey Mouse and similar brands on Disney Channel Asia. In Disney movies and magazines, they are also exposed to Western traditions including Halloween and St Valentine's Day. Disney is not deliberately trying to change cultures in other countries, but this is often the result. The apps which are bundled up with an Apple iPhone promote Western culture in more subtle ways: St Valentine's Day (named after a Christian saint) may be mentioned in a calendar, for instance.

The opposing view to this is that the actions of TNCs cannot be described as cultural imperialism because businesses operate independently of governments. US-based TNCs cannot be accused of spreading US 'propaganda' because they are motivated by profits, not politics. The guiding principle of these firms is simply to build market share on a global scale. In doing so, they may well introduce cultural changes to places. They can only do so, however, with the consent of their customers; 'imperialism' sounds like a misnomer for a cultural process which does not use force.

■ The glocalization of branded commodities

The practice of glocalization involves TNCs adapting their products for different markets to take account of local variations in tastes, customs and laws. This strategy developed originally from the need of some TNCs to source parts and ingredients locally when establishing branch plants overseas. SABMiller, a major TNC, uses cassava to brew beer in Africa, for instance; this cuts the cost of importing barley (which is used in other world regions).

Glocalization makes business sense too because of geographical variations in:

- *people's tastes* European TNC GlaxoSmithKline rebranded its energy drink Lucozade for the Chinese market with a more intense flavour. Working in partnership with Uni-President China Holdings, the product's new local name translates as 'excellent suitable glucose'.
- *religion and culture* Domino's Pizza offers only vegetarian food in India's Hindu neighbourhoods; MTV avoids showing overtly sexual music videos on its Middle Eastern channel; early beer advertising in West Africa, shown in Figure 5.20, made use of local gender stereotypes.
- *laws* The driving seat should be positioned differently for cars sold in US and UK markets.

Figure 5.20 Early 'glocalized' advertising of Guinness beer in Sierra Leone, 1968

- *local interest* Reality TV shows, such as *The X-Factor*, gain larger audiences if they are re-filmed using local people in different countries.

Driving this attention to detail is the sheer size of emerging markets.

Strengthening Latin American, Asian, African and Middle Eastern economies are home to growing numbers of cash-rich young people in places previously sidelined by TNCs. Leading global brands want to maximize their sales chances. Glocalization has therefore become a critically important economic, political and cultural strategy that informs and defines certain kinds of companies' actions in the global marketplace (Table 5.6).

At its most effective, glocalization can rejuvenate an entire industry. Global firms can mine new local territories to create fresh glocalized music, food or fashion ideas that are fed back in turn to core markets. (Spiderman India has been adopted now by US Spiderman fans, for instance.) Japanese, Indian, Korean and Nigerian influences, among others, increasingly drive innovation in global creative industries. Film, music and food industries have all thrived by mixing together Asian, South American and African influences with European and American ideas.

Table 5.6 Examples of glocalization

McDonald's Corporation	MTV (Music Television)	The Walt Disney Company
By 2015, McDonald's had established 35,000 restaurants in 119 countries. In India, the challenge for McDonald's has been to cater for Hindus and Sikhs, who are traditionally vegetarian, and also Muslims, who do not eat pork. Chicken burgers are served alongside the McVeggie and McSpicy Paneer (an Indian cheese patty). In 2012, McDonald's opened a vegetarian restaurant for Sikh pilgrims visiting Amritsar, home of the Golden Temple. The success of these glocalizing strategies owes much to the local knowledge of Connaught Plaza Restaurants which works with McDonald's in this joint venture (see page 21).	MTV Networks use a 360-degree strategy involving 'full spectrum' marketing to maximize its global audience. It has grown over time through acquisitions and mergers with existing companies, as well as by setting up new regional service providers such as MTV Base (available since 2005 to around 50 million viewers across 48 countries in sub-Saharan Africa via satellite). In 2008, new channel MTV Arabia began broadcasting to Egypt, Saudi Arabia and Dubai. Two-thirds of the Arab world is younger than 30 – and many of these young people are fans of cutting-edge music, especially hip-hop, which MTV Arabia now specializes in broadcasting.	*Roadside Romeo* (2008) was the first film that Disney made inside India. It was aimed at local audiences and used home-grown animation put together by Tata Elxsi's Visual Computing Labs (VCL) unit. A co-production with India's Yash Raj studios, this film tells the story of a dog living in Mumbai. Disney acquired Marvel in 2009, gaining the rights to superhero characters that have sometimes been glocalized. *Spiderman India* is an example. In a story made for Indian children, Mumbai teenager Pavitr Prabhakar is given superpowers by a mystic being. The story is different from the version children in many other countries are familiar with.
'We have to keep our ears to the ground to know what the local customer desires … [it's] key to our worldwide functioning.' (Amit Jatia, MD of McDonald's India West and South)	'We will respect our audience's culture and upbringing without diluting the essence of MTV.' (Bhavneet Singh, MD of MTV Emerging Markets)	'There is great interest and pride in local culture. Even though technology is breaking down borders, we're not seeing homogeneity of cultures.' (Bob Iger, Disney CEO)

Could TNCs using glocalization strategies ever be accused of cultural imperialism? At first glance, the sensitivity of these companies to local cultures, norms and needs appears to weaken the case. However, there is a counter-argument to consider, as follows:

- It might be said that a McDonald's burger sold in India is still a vehicle for the Americanization of Asia, irrespective of the fact that extra spice is added and a beef substitute used (in deference to Hindu taste and belief). Consider also the 'gate-keeping' role that MTV's international executives may play in selecting local talent to slot in among the majority Anglo-American output of their regional satellite stations. The most favoured local artists will be those who complement mainstream commercial programming. More diverse indigenous musical expressions are more likely to be left languishing in relative obscurity on the sidelines.
- The glocalized nature of goods and services sold at local markets by TNCs must also not blind us to the fact that a more fundamental sweeping change is still occurring, namely commercialization of food, fashion and drink, among many other elements of people's lifestyles. Globally dominant firms are responsible for spreading the consumerism at a planetary level. The twentieth-century philosopher Antonio Gramsci used the term **hegemonic power** to describe influence on this scale. Powerful TNCs are shaping a new, popular common

> **Keyword definition**
> **Hegemonic power** The ability of a powerful state or player to influence outcomes without reverting to 'hard power' tactics such as military force. Instead, control is exercised using a range of 'soft' strategies of persuasion, including diplomacy, aid and the work of the media and educational institutions.

sense that sees their products as valued and desired by people living in newly emerging economies such as Brazil or India. Long-established un-commercial forms of leisure, play and entertainment are rapidly giving way to consumer aspirations among growing sections of the populations living in these places.
- Arguably, powerful TNCs are behaving in a coercive and imperialistic way: they use clever advertising techniques to manipulate the desires and aspirations of people on a global scale. Noam Chomsky calls this 'the manufacturing of consent' (Figure 5.21). According to this viewpoint, any impression that local people are equal partners in a 'cultural conversation' taking place with TNCs is false. Global food and drink conglomerates like McDonald's and Diageo merely pay lip service to local cultures by 'tweaking' the taste of their generic products, or the slant of their advertisements, in order to drive sales higher in differing markets.

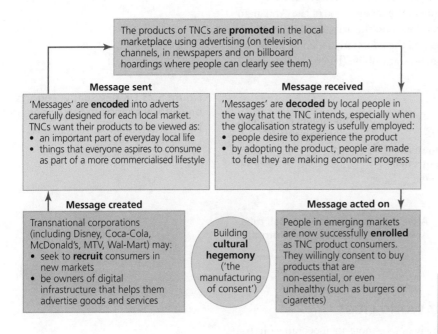

Figure 5.21 How TNCs use the power of persuasion to 'enroll' communities as consumers of their products in different local contexts around the world

■ Global interactions and cultural hybridity

Glocalized products are just one example of **cultural hybridity** resulting from the 'mash-up' of global and local cultural elements. Cultural outcomes of global interactions – as experienced at the local level – are often complex.

When local communities interact with different global flows it may result in aspects of their culture being changed; but they do not necessarily lose their own unique identity as a consequence. Instead a hybrid culture results that combines elements of the local/traditional with modern/global characteristics.

■ Increased cultural hybridity among rainforest tribes

Rainforest tribes are often portrayed as being among the world's least developed – and least globalized – societies. Amazonian and Papua New Guinea's tropical rainforest people are among the world's last isolated groups of indigenous people. These ethnic groups have occupied the place where they live for thousands of years without interruption. More members of rainforest tribes are becoming aware of Western culture and lifestyles, however:

- Owing to the tropical climate, indigenous rainforest people traditionally wore little in the way of clothing. Today, many Amazonians and New Guineans instead are wearing modern, Westernized clothing such as the T-shirt, or have adopted branded products like Coca-Cola (Figure 5.22).

> **Keyword definition**
>
> **Cultural hybridity** When a new culture develops, whose traits combine two or more different sets of influences.

Figure 5.22 Indigenous people are increasingly exposed to global culture, resulting in cultural hybridity

- This does not mean that their traditional culture has been lost, though. Modern brands can be adopted without important customs being abandoned.

In 2014, Brazilian news reports showed how members of the Kuikuro tribe were enjoying the country's hosting of the football World Cup. The Kuikuro are a small indigenous group of fewer than 600 people who live in Brazil's Xingu region.

- One anthropologist who was working with the Kuikuro tribe saw they were really enjoying watching matches on television; Chief Afukaká was reported to be a great football fan.
- Some 15 Kuikuro Indians obtained tickets to watch a football match between Russia and South Korea. They gained their tickets via an allocation for the National Indian Foundation (FUNAI) and attended the match in the central western Brazilian city of Cuiaba.

Not everyone views cultural change and hybridity among indigenous tribes positively, however. The traditional culture of the Jarawa tribe of the Andaman Islands has been affected greatly by global flows and the arrival of large numbers of tourists in particular. Cultural dilution and the adoption of alcohol are viewed by some people as grave threats to the Jarawa's way of life (Figure 5.23).

Figure 5.23 UK newspaper reports on cultural change in the Andaman Islands

Source: Gill Miller

The cultural landscape of world cities

Are the world's cities becoming increasingly homogenous in appearance? In some ways, the cultural landscape of many major cities – including São Paulo, New York, Singapore, Mumbai, Tokyo and Shanghai – is increasingly homogenous:

- Most large cities contain what critics call 'cloned' retailing districts: many modern steel-and-glass shopping malls resemble one another very strongly, thanks in part to the ubiquity of global brands including McDonald's and Starbucks.
- The modern architecture found in financial and high-class housing districts often appears more 'global' than 'local' in style. The Pudong financial district in Shanghai is a typical global **financescape**. Similarities between places arise in part due to the transnational work of leading global architects and architecture firms. For instance, the Uruguayan architect Rafael Viñoly has designed striking skyscraper structures for numerous world cities including London, New York and Los Angles (Viñoly's designs are controversial: his 37-storey 'Walkie-Talkie' building in London gained notoriety when its concave shape channelled the sun's rays into a concentrated beam of light which melted the pavement and cars below!)
- Inter-urban competition drives the adoption of tall modern designs: for many years, world cities have competed against one another to have the tallest building in the world (in 2016, the title was held by Burj Khalifa in Dubai). Planners believe that imposing modern designs send a signal that their city is a modern, world-class place to do business. Impressing others with your architecture helps to send a message that your city is a powerful global hub (Unit 4.11, page 10).
- The power of TNCs to project their brands and global advertising messages in urban environments also drives change: many of the world's tallest buildings have corporate logos mounted at the top (Figure 5.24).

> **Keyword definition**
> **Financescape** A modern landscape of tower blocks and offices that incorporates state-of-the-art architecture, and which is usually designed to impress by reaching greater heights than the surrounding district.

The global diffusion of new architectural forms is part of a much wider shrinking world effect, which can lead to landscape changes: the global transmission of ideas, information and people inevitably leads to the spread of a global culture, as this unit has showed repeatedly. The increasingly homogenous visual appearance of urban areas is another manifestation of this global culture.

The importance of demographic processes – including population growth, rising affluence and urbanization in emerging economies – should not be underestimated either. During the past decade, more than 1 billion people have moved from rural to urban areas across the world. Factor in natural increase and you can see why an unprecedented amount of new urban development has been taking place – China alone has recently had to build 200 'instant cities' from scratch to hold 1 million people each. Faced with a housing crisis, city planners throughout the developing world have adopted similar high-rise solutions.

■ Traditional and hybrid city landscapes

Despite the many similarities that exist, all cities are not the same, of course. This is because of regional variations in cultures and laws that help to determine what can and cannot be built (or demolished). The physical environment matters too – tall towers might not be advisable in areas where certain natural hazard risks exist.

Many older cities have a hybrid appearance which blends new 'global' features with surviving traditional and historic buildings and districts. This is called a 'palimpsest' landscape. This means it contains traces of both recent and far older societies, activities and cultures. In successive historical periods, new 'layers' of development were laid down like a blanket; but these blankets contained holes and did not cover everything that had come before.

- Medieval churches and houses can be seen in many older European towns and cities, despite being surrounded by more recent housing and offices. Strict laws protect these historic buildings.
- Many cities in Asia, Africa and Latin America have ancient districts that have been awarded World Heritage Site (WHS) status by the United Nations Educational, Scientific and Cultural Organization (UNESCO), whose aim is the 'preservation and promotion of the common heritage of humanity'. Cairo's Old Town, where Sultan Hassan Mosque is situated, has WHS status, for instance.

Figure 5.24 'Branded' buildings in a financescape: this is Hong Kong, but it could be almost any world city

However, population pressure on Cairo and other developing world cities is even greater than that experienced by Europe in the past. Some megacities like São Paulo and Lagos now receive half a million new arrivals annually. Shenzhen in China has grown from a small market and fishing town of fewer than 300,000 inhabitants to a sprawling metropolis of 20 million people in just 35 years. Growth is so rapid that protection of older districts becomes a challenge.

The impact of global diaspora growth on cultural diversity and identity

Another aspect of cultural globalization to consider is the worldwide scattering or dispersal of a particular nation's migrant population and their descendants. This is called diaspora. Over time, each diaspora's cultural traits are preserved (albeit in a modified form) and connections are maintained between groups of peoples with common ancestry living in different territories. Famous examples include:

- *the Jewish diaspora* This first historical diaspora dates from nearby 2000 years ago when the Jewish population was forced to disperse, following the Roman conquest of Palestine. Today, the state of Israel serves as the hub for a global Jewish population that has notable concentrations in Europe, North America and Russia.
- *the 'Black Atlantic' diaspora* This has been described by writer Paul Gilroy as a 'transnational culture' built on the movements of people of African descent to Europe, the Caribbean and the Americas. A shared, spatially dislocated history of slavery originally helped to shape this group's identity. Today, international connectivity is maintained through tourism and cultural exchanges across the Atlantic that are well exemplified by an international Black music scene that has given the world jazz, Jimi Hendrix, reggae and hip-hop.
- *the new Polish diaspora* Estimates vary from 1 million to 3 million for the number of young people who left Poland after it joined the European Union in 2004. These recent émigrés join a larger and well-established mass of around 20 million people of Polish descent living abroad – some of whom are the children of Second World War refugees.
- *the Chinese diaspora* The neighbouring countries of Indonesia, Thailand and Malaysia, along with far-flung places such as the UK and France, have significant Chinese populations (Figure 5.25). In many world cities clearly delimited 'Chinatown' districts exist. A thousand years of seafaring trade gives this diaspora a long history. The arrival of Chinese TNCs in Africa has brought further diaspora growth in recent years.

There are many more significant examples that can be researched, including French, Indian, Italian, Mexican, Brazilian, Nigerian or Malay diasporas. Key to a successful analysis is recognizing the cultural variations that arise over time between different parts of the diaspora. This is due to the cultural hybridity than can develop in each local context as migrants and their descendants adopt elements of the 'host' country's culture.

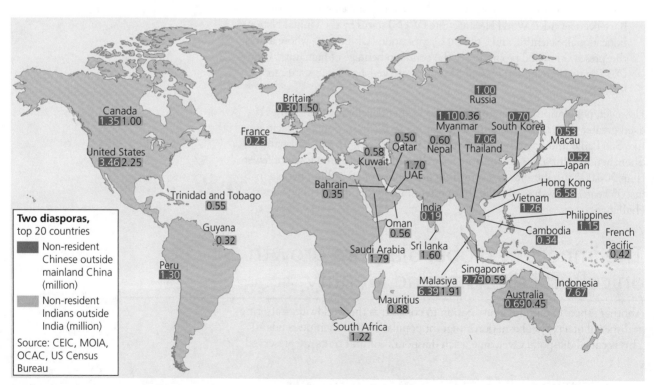

Figure 5.25 Indian and Chinese non-resident citizens living abroad in selected countries in 2011 (millions more people of Indian or Chinese descent are also scattered worldwide)

The Irish, Scottish and Welsh diasporas

The UK's 'Celtic fringes' have all birthed significant global diasporas despite these nations' relatively small population sizes. For instance, Ireland is home to just 4 million people, yet over 70 million individuals living worldwide claim Irish ancestry. In the USA alone, 30 million people believe themselves tied to Irish bloodlines, following mass emigration from Ireland during the nineteenth and early twentieth centuries.

Online ancestry websites enable people living all over the world to trace their roots back to Scotland: this is an interesting way in which technology has influenced global interactions. People who discover they have roots in another country may become curious and decide to visit as tourists, thus giving rise to another global flow. The Scottish tourist industry estimates around 3 million visitors from overseas annually, many of whom count themselves as diaspora members. GlobalScot is a website run by government-funded Scottish Enterprise that actively encourages members of the Scottish diaspora to network economically with one another.

Sensing business opportunity, a large part of Scotland's tourist industry plays up to the expectations of tourists with Scottish roots who want to see 'tartan and bagpipes'. However, many young people living in Scotland object to this fetishization of the past: they fear their nation risks becoming culturally 'fossilized'.

CASE STUDY

THE INDIAN DIASPORA

The Indian diaspora is one of the world's largest, numbering 28 million in 2016. People of Indian citizenship or descent live in almost every part of the world. Important features of the pattern are as follows:

- It numbers more than 1 million in each of 8 countries. The largest concentrations are in the USA, UK, Malaysia, Sri Lanka, South Africa and the Middle East.
- Some 22 countries have concentrations of at least 100,000 ethnic Indians.

An important distinction exists between non-resident Indians (NRIs) and persons of Indian origin (PIOs). This is shown in Table 5.7.

Table 5.7 Comparing non-resident Indians (NRIs) and persons of Indian origin (PIOs)

NRI facts	• Around 16 million members of the diaspora are NRIs (India has more of its actual citizens living abroad than any other country).
	• They are mostly economic migrants; many are highly skilled and young, reflecting India's current demographic characteristics. The country's median age is just 26 and it has the second largest population in the world: it is therefore unsurprising that many of the world's young economic migrants have travelled from India.
	• Many have travelled to work in elite occupations such as technology and medicine. Many skilled Indian migrants have travelled to the USA since the 1990s; the NRI population in the USA is more skilled and highly paid than any other migrant community living there.
	• In contrast, many Indians working in the Gulf States are low-skilled and poorly paid (see Unit 4.2, page 16).
PIO facts	• The remaining 12 million members of the diaspora are PIOs.
	• The exact number is hard to estimate as many fourth- or fifth-generation descendants of Indian migrants may no longer identify themselves as being ethnically Indian. Many have a multiple identity – this could be someone with an Indian grandmother and three Irish grandparents, for instance!
	• Different state governments collect information about their citizens' ethnicity in varying ways and may not even ask questions about people's cultural backgrounds as part of their national census.
	• There are many good reasons why we should question the validity and reliability of PIO diaspora data (and the same is true for any other diaspora you might decide to research).

In the context of your course, there are several interesting points to consider when studying the Indian diaspora:

- The diaspora contributes to India's growing global power and influence (Unit 4.1). India's government views its diaspora as an important human resource helping to build Indian soft power overseas. In 2015, India's Prime Minister Narender Modi found time to visit UK Indian diaspora communities while making a state visit to London. The diaspora population plays an important role in supporting the enduring friendship between UK and India – a relationship that both countries arguably benefit from.

- The remittance flows that return to India are the world's largest. They were valued at US$70 billion in 2015: this is a highly significant global financial flow (Unit 4.2).

The diaspora's characteristics vary from country to country: many different hybrid cultures have been created. The degree to which PIOs have preserved their traditional culture and have partly assimilated into the society they live in varies according to context and for many reasons:

- Local attitudes towards mixing may create constraints on integration.
- More recently established diaspora communities have had less time to become modified.
- Indian NRIs and PIOs belong to different Hindu, Muslim and Sikh communities: each of these religious groups may interact with other local communities in different ways, giving rise to an even greater range of hybrid cultures. One example of a hybrid culture is the British Indian Sikh community (now the largest Sikh community outside India). Each wave of Sikh migrants to the UK has brought its own cultural beliefs with it and yet managed to integrate itself within British society while retaining its distinct identity; this has led to the formation of a unique British Sikh identity.

■ KNOWLEDGE CHECKLIST

- The concepts of culture, cultural traits, ethnicity and identity
- Evidence that diversity is narrowing globally to create a 'global culture'
- Evidence that cultural diversity is increasing in some states and city contexts
- The diffusion of cultural traits and how this affects places
- Cultural imperialism and how this affects places
- Examples of the glocalization of branded commodities
- The concept of cultural hybridity
- Cultural landscape changes and the increasing homogeneity of built environments at a worldwide scale
- The importance of diasporas and their influence on cultural diversity and identity at global and local scales, using a case study of a global diaspora population (Indian diaspora)

EVALUATION, SYNTHESIS AND SKILLS (ESK) SUMMARY

- How views may differ on the desirability of cultural change at global, national and local scales
- How the evidence for cultural change is complex, difficult to collect and sometimes contradictory

EXAM FOCUS

MIND-MAPPING USING THE GEOGRAPHY CONCEPTS

The course Geography Concepts were introduced on page vii and also feature throughout Units 4.1 and 4.2.

Use ideas from Units 5.1 and 5.2 to add extra detail to the mind map below (based around the Geography Concepts) in order to consolidate your understanding of human development and diversity.

To what extent is there a global culture?

- **Spatial interactions**: Different global flows have fostered cultural change and exchange between places over time: this has led to the global diffusion of certain traits like the English language
- **Place**: Glocal culture combines with the culture of local places to create new hybrid cultures: no two places are exactly the same; McDonald's menus vary from place to place
- **Scale**: While it may be true that there are aspects of a common culture at the global scale, cities and towns often have great cultural diversity and many different ethnic groups – rather than one single culture
- **Power**: If there is a global culture, where does it come from? Which powerful countries have the greatest imprint on the clothes we wear and music we listen to? Who has most soft power?
- **Possibility**: Will global culture change as new superpowers like India and China gain greater power and influence over the world economy and the products and services we consume?
- **Processes**: Many important processes affect how cultures change and evolve, including glocalization, cultural imperialism, the melting pot effect and Westernization

Spectrum of importance

Below is a sample part (b) exam-style question and a list of factors that could be used to answer the question (add other factors you think are important). Use your own knowledge and the information in this section to reach a judgement about the importance of these factors to the question posed. Write numbers on the spectrum (right) to indicate their relative importance. Having done this, write a brief justification of your placements, explaining why some of these factors are more important than others. The resulting diagram could form the basis of an essay plan.

Don't forget to make use of the Geography Concepts as part of your evaluation.

Evaluate the role that different factors have played in the growth of a global culture. (16 marks)

Possible factors:	Your ranking
Colonialism	
TNCs	
Government rules	
Migration	
The internet	
Transport technology	
Diaspora growth	
Hollywood and the media	

5.3 The power of places to resist or accept change

In the view of many political and economic commentators, barriers to global interactions have risen not fallen in recent years. A series of geopolitical shocks has shaken the faith of hyperglobalizers (Table 5.8). It is becoming increasingly commonplace to read media reports declaring that globalization is 'in retreat'. In Unit 4.2, we saw that trade flows have lessened in value in recent years. Yet we also know that global data flows continue to accelerate. Evidence of so-called **de-globalization** is therefore highly contradictory and no clear case emerges.

Table 5.8 Recent 'shocks' to globalization

2001	Attack on US World Trade Center	Al-Qaeda's violent act signalled the start of the so-called 'war on terror'. Daesh has subsequently emerged as the major player in an ongoing conflict that has brought misery to millions in the Middle East. Daesh attacks on Western cities and tourists appear calculated to disrupt 'business as usual' globalization.
2008	Global Financial Crisis (GFC)	The GFC was rooted in US and EU money markets. High-risk lending by banks eventually undermined the entire world economy. Global GDP fell for the first time since 1945, also triggering the Eurozone crisis.
2011	Arab Spring	At first, popular uprisings against North African and Middle Eastern dictators appeared to signal political progress. But the global community failed to avert a crisis in Syria.
2014	Russian annexation of Crimea	The protection of the human rights of ethnic Russians living there was the excuse used by Russia when it invaded Ukraine. The international community condemned this but not steps were taken by Russia to abondon Crimea.
2016	UK votes to leave the EU	Could 'Brexit' spark the eventual disintegration of the EU if other populations demand a referendum too?

Power to take action to resist any or all of these aspects of globalization is invested in many different stakeholders at a range of geographic scales and positions of responsibility. The heads of state of several countries, including Venezuela, Cuba and North Korea, are vocal opponents of the 'new world order'. So too are influential non-governmental organizations (NGOs) such as Amnesty International, Greenpeace and Christian Aid (Figure 5.26). NGOs are part of **civil society,** along with private citizens, many of whom strongly believe that the poorly regulated actions of globalization's most powerful economic players – such as international banks and food companies – should be opposed, or at least have their excesses curbed and damaging effects ameliorated.

Keyword definitions

Civil society Any organization or movement that works in the area between the household, the private sector and the state to negotiate matters of public concern. Civil society includes non-governmental organizations (NGOs), community groups, trade unions, academic institutions and faith-based organizations.

De-globalization A reduction in the intensity of some global interactions or the introduction of new barriers to some global flows.

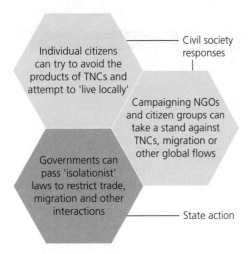

Figure 5.26 Varying scales of resistance to global interactions

Local and civil society resistance to global interactions

There are many instances of local and civil society resistance to globalization (the attitudes of governments are explored later in this unit). Populist anti-globalization and **nationalist** movements are found in every country in the world although their foci are highly variable. Different countries are affected by global flows in varying ways: migration, cultural change, sovereignty loss, deindustrialization, outsourcing and environmental degradation are just some of the processes that can trigger resistance movements in certain place contexts. Figure 5.27 shows two contrasting rationales for resistance drawn from opposing ends of the political spectrum.

There is a clear and *increasing* correlation between the growing interconnectedness of states and the desire of many citizens of those states to sever ties with other places. The paradox of globalization is that while it promotes a global way of living, it simultaneously ignites fears and concerns that local identities may be eroded and lost.

> **Keyword definitions**
>
> **Nationalist** A political movement focused on national independence or the abandonment of policies that are viewed by some people as a threat to national sovereignty or national culture.

Sustainability and justice	Nationalism and sovereignty
• Some resistance movements are anchored in concerns with social, economic and environmental justice; unethical aspects of globalization are a threat to sustainable development. • Demonstrators worry about the impact of global interactions on the lives of their own communities *and other people too*. Issues include the exploitation of agribusiness labourers and 'sweatshop' workers, and the local and global environmental damage done by globalized economic activity, including global warming.	• Other resistance movements are anchored in renewed nationalism. This often-powerful 'grass roots' force can be reinforced at an institutional level by the media and political parties. • In many places, nationalism is on the rise. The shrinking world, so the argument goes, has caused chaos for national life and culture due to new and sometimes unchecked flows of people, information and ideas. As a result, people's *own community identity* is threatened and *their own state's sovereignty* (independence) has been jeapordized.

Figure 5.27 Contrasting civil society rationales for resistance to global interactions

■ The rejection of globalized production

TNCs have often been the target for civil society sustainability and social justice campaigns. In recent years, many companies have attracted criticism for a range of reasons (Table 5.9). However, as Units 4.2 and 5.2 demonstrate, TNCs may also bring benefits to places, especially when social responsibility frameworks are used. This is why civil society movements targeted against TNCs do not always gain wider popular support.

> **PPPPSS CONCEPTS**
>
> Think about the way anti-globalization protests are often focused on one particular process – such as migration or trade – but not others, such as internet and social network use. There is a spectrum of resistance to globalization.

Table 5.9 Reasons for civil society opposition against some TNCs and their production and trade networks

The case against some TNCs
1 *Growing global wealth divide* The actions of TNCs are building a widening global wealth gap between the very richest and most poor countries. By selectively investing in certain regions (Southeast Asia) while largely by-passing others (Saharan Africa), TNCs are active agents in creating a new geography of 'haves' and 'have-nots'. In some exceptional cases, TNC control – and accountability – is found to be invested in the hands of a few easily identifiable individuals. Wal-Mart famously remains under the influence of the Walton family, whose combined inherited wealth is estimated to be around US$80 billion. The Waltons are both majority shareholders and company directors.
2 *Social harm* Hundreds of millions of the world's poorest people have become merged into a hard-labouring global proletariat spanning all sectors of industry. Foreign Direct Investment (FDI) does not always result in a step-change for economic activity in developing countries. If the TNC employs local firms to help it carry out its work – for example, in the sourcing of raw materials – then more wealth may be generated, but this does not always happen.
3 *Environmental degradation* Critics of global economic expansion hold its agents responsible for serial polluting and destruction of fragile local habitats as well as accelerated emissions of greenhouse gases and runaway global climate change. One of the most notorious cases of this occurred on 2 December 1984, when a deadly fog of poisonous gas was emitted from a pesticide plant in Bhopal, India, owned by the American TNC Union Carbide. A lethal plume of methyl isocyanate resulted in the deaths of thousands of Indians living close to the plant in Bhopal.
4 *Tax avoidance* TNCs including Starbucks and Google have been accused of not paying full corporation taxes in the countries where they operate, through transfer pricing and tax concessions (see Unit 6.1). This means that governments find it harder to raise revenues, provide services and respond to the demands of local people. Citizens have protested against this (Figure 5.30).
5 *Cultural imperialism* Threats to cultural diversity include the loss of world languages. Critics of some TNCs say they have established a hegemony that encourages greater cultural conformity between different localities through the mass consumption of branded products (see Unit 5.2).

Political demonstrations and protests against globalization became more commonplace during the late 1990s and 'noughties', often orchestrated by private citizens interfacing with new technology such as Facebook or Twitter. Key oppositional moments in recent times include:

- *Seattle World Trade Organization conference, 1999* Around 40,000 protestors flooded the streets and voiced disapproval of WTO guidelines for global trade that critics have argued prevent poorer nations from 'playing on a level playing field'. Full-scale rioting resulted in a state of emergency being declared in Seattle.
- *Cancun, Mexico world trade talks, 2003* South Korean farmer Lee Kyung-hae died after purposefully stabbing himself in the chest in protest against a continuing lack of reform for WTO global trade laws. Around 5,000 angry protestors were later held back by police.
- *Paris climate change conference, 2015* Many hundreds of campaigners were detained by police after activists, some dressed as polar bears and penguins, took to the streets of Paris while the world's political leaders attempted to create new binding agreement to tackle global climate change.

■ Civil society protests against energy TNCs in Canada

Canada is home to six groups of indigenous people known as the First Nations. Their occupation of the land long pre-dates the arrival of Europeans. Some First Nations people of the Mackenzie and Yukon River Basins oppose the attempts of global oil companies – including Shell, ExxonMobil and Imperial Oil – to exploit the natural resources of their region (an area of boreal forest and tundra). The Dene residents of the Sahtu Region have already experienced negative impacts of globalization and petroleum development near the settlement of Norman Wells. More than 200 million barrels of conventional oil have been extracted there since 1920. Particular concerns include:

- the death of trout and other fish in oil-polluted lakes (a lifestyle based around subsistence fishing, hunting and trapping is fundamental to the Dene's cultural identity)
- the effects of alcohol and drugs (brought by oil workers) on the behaviour of young Dene people.

Oil TNCs are now exploring the surrounding Canol shale and assessing its potential for shale oil (Figure 5.29). Shale 'fracking' (hydraulic fracturing) in other places has been linked with water pollution. In east Canada, the Elsipogtog First Nation were involved in a violent protest against fracking: police vehicles were set on fire after tear gas and rubber bullets were fired at protesters.

Figure 5.28 Protests outside Starbucks

Figure 5.29 Unconventional fossil fuel resources in Canada

■ Civil society protests against McDonald's in India

Despite McDonald's use of glocalization to gain acceptance in different markets including India (see Unit 5.2, page 62), the company attracts harsh criticism on the grounds that it is responsible for cultural homogenization on a global scale. There is concern that local cultures are being 'extinguished' as a result of the global 'roll-out' of uniform (albeit glocalized) services, such as McDonald's restaurants – some people even term it 'McDonaldization'. Globalization skeptics are deeply concerned that local cooking traditions will be lost to the onslaught of fast-food menus (often written in English too, thereby accelerating the decline of local languages).

- In 2012, McDonald's took the step of opening two entirely meat-free vegetarian restaurants in India. Religious pilgrims to two of India's most sacred spiritual sites found the golden arches of McDonald's waiting there for them.
- These vegetarian outlets were developed in Amritsar, home of the Golden Temple, the holiest site of India's minority Sikh faith, and the town of Katra, the base for Hindus visiting the mountain shrine of Vaishno Devi, the second busiest pilgrimage spot in India.
- The move was not popular with some people, however. The Hindu nationalist group Swadeshi Jagran Manch, a branch of the influential Rashtriya Swayamsevak Sangh (RSS), opposed the arrival of McDonald's. 'It's an attempt not only to make money but also to deliberately humiliate Hindus. It is an organization associated with cow slaughter. If we make an announcement that they're slaughtering cows, people won't eat there. We are definitely going to fight it', a spokesperson told one newspaper in 2012.
- However, McDonald's has since gone from strength to strength in India, serving over 300 million customers in 2015, including those in Amritsar. Burger King, KFC and Dunkin' Donuts all have a growing presence in India too.

■ Local sourcing of food and goods

The 'local sourcing' movement describes the practice of buying goods and services solely from local area suppliers, thereby boycotting the use of extended global supply networks such as those favoured by most supermarkets. The clearest environmental benefit is believed to be a markedly reduced number of food miles for agricultural products. Providing stimulus for local production has other merits too. Greater regionalized agricultural activity can improve national food security in an era when climate change and growing global demand put mounting pressure on the world's food supplies. In developed nations, greater local sourcing of food can also boost employment in depopulated rural areas.

A cursory look at any supermarket aisle will yield no shortage of instances where imported food, drink and consumer goods have travelled truly excessive distances. Freshly picked vegetables like asparagus are routinely transported into Europe by air from South America. UK supermarket Sainsbury's sources charcoal briquettes for barbeques from South Africa. Setting light in a British garden to wood that has been burned once already – having also virtually circumnavigated the globe in a carbon-emitting container ship vessel – surely ranks high on any scale measuring environmental folly.

Local sourcing wins plaudits from governments and is promoted enthusiastically by those businesses who have declared their commitment to carbon footprint reduction. However, some foreign agricultural imports into the UK – notably Spanish and North African tomatoes and flowers – may actually do less environmental harm than some local agricultural systems that are reliant on energy-hungry heated greenhouses (Figure 5.30). Blaming globalization for increased carbon emissions can sometimes be a misreading of a more fundamental issue. After all, globalization is not directly responsible for British people wanting to buy large amounts of flowers in the winter month of February. The root cause instead is commercialization of the Valentine's Day festival. This drives demand for flowers that can be met only by flying equatorial and southern hemisphere flowers to market in the UK or through greater local use of heated greenhouses.

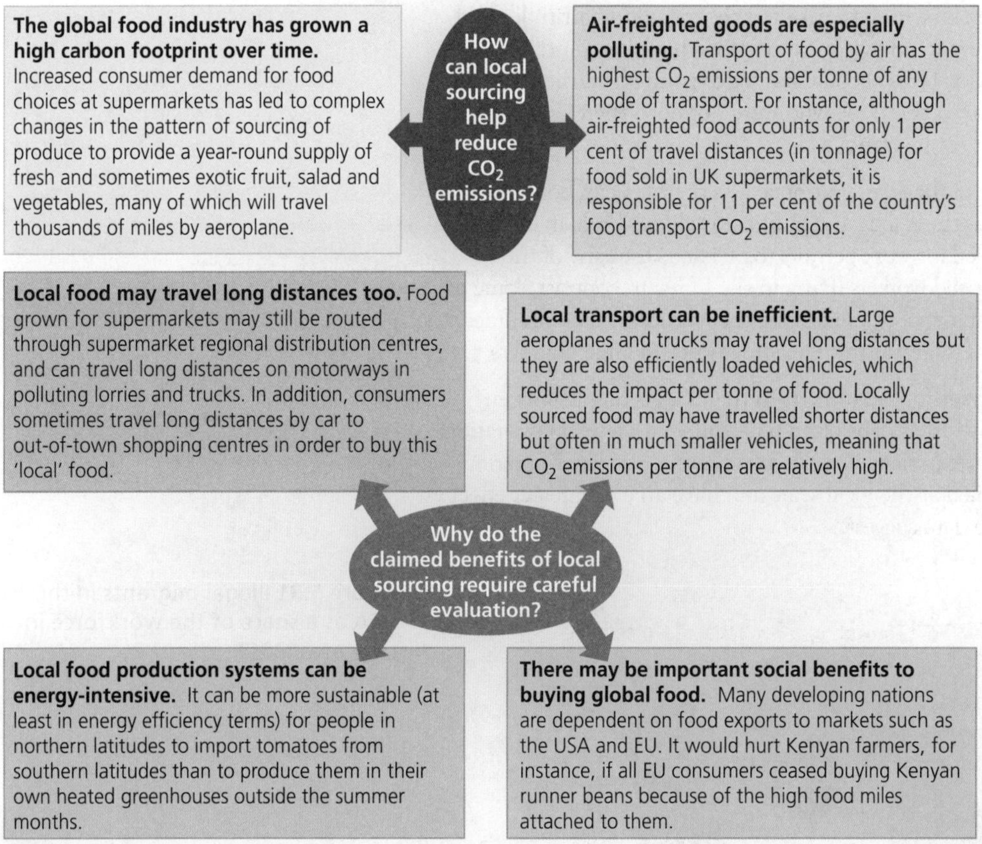

5.3 The power of places to resist or accept change

The global food industry has grown a high carbon footprint over time. Increased consumer demand for food choices at supermarkets has led to complex changes in the pattern of sourcing of produce to provide a year-round supply of fresh and sometimes exotic fruit, salad and vegetables, many of which will travel thousands of miles by aeroplane.

How can local sourcing help reduce CO_2 emissions?

Air-freighted goods are especially polluting. Transport of food by air has the highest CO_2 emissions per tonne of any mode of transport. For instance, although air-freighted food accounts for only 1 per cent of travel distances (in tonnage) for food sold in UK supermarkets, it is responsible for 11 per cent of the country's food transport CO_2 emissions.

Local food may travel long distances too. Food grown for supermarkets may still be routed through supermarket regional distribution centres, and can travel long distances on motorways in polluting lorries and trucks. In addition, consumers sometimes travel long distances by car to out-of-town shopping centres in order to buy this 'local' food.

Local transport can be inefficient. Large aeroplanes and trucks may travel long distances but they are also efficiently loaded vehicles, which reduces the impact per tonne of food. Locally sourced food may have travelled shorter distances but often in much smaller vehicles, meaning that CO_2 emissions per tonne are relatively high.

Why do the claimed benefits of local sourcing require careful evaluation?

Local food production systems can be energy-intensive. It can be more sustainable (at least in energy efficiency terms) for people in northern latitudes to import tomatoes from southern latitudes than to produce them in their own heated greenhouses outside the summer months.

There may be important social benefits to buying global food. Many developing nations are dependent on food exports to markets such as the USA and EU. It would hurt Kenyan farmers, for instance, if all EU consumers ceased buying Kenyan runner beans because of the high food miles attached to them.

Figure 5.30 Resisting globalization through the local sourcing of food

Opting to buy fewer non-essential goods such as flowers in the first place – irrespective of their local or global origin – would be an even more effective way for wealthy societies to mitigate climate change than local sourcing (but would also be enormously damaging for global trade and the livelihoods of thousands of people in poor countries like Kenya).

Rise of anti-immigration movements

The rationale for retreating from globalization is rooted, for some people, in the valid concern that national cultural identity is being changed by global interactions. Migration in particular creates political tensions due to differing perceptions of, and viewpoints on, the cultural changes it brings.

For EU nations, the renewal of nationalism is linked with a broader debate about 'loss of sovereignty'. A large proportion of citizens in each EU country would like to end the freedom of movement brought by the Schengen Agreement (see Unit 4.3, page 30). They believe too much immigration has been allowed to take place.

- In some EU states, nationalist parties, such as France's Front National, command significant support. Nationalist parties often oppose immigration; some reject multiculturalism entirely and embrace fascism openly.
- When a majority of UK voters chose to leave the EU in a referendum on membership in 2016, immigration was the most important issue influencing how people voted. Thirty per cent of London's 8 million residents were born in another country; some British citizens judge the scale and rate of cultural change in the UK to have been too great.
- In France in 2015, staff of the satirical magazine *Charlie Hebdo* were killed by gunmen of Algerian descent. The murderers said their Islamic faith had been mocked. Extreme events such as these are still rare but demonstrate tensions in multicultural Europe, which may ultimately threaten the survival of free movement of people.

Differing perspectives on immigration in the USA

The USA is an interesting and important example, given the sheer volume of migrants and their descendants who live there. Between 1900 and 1920 alone,

24 million new arrivals were registered, thanks to the 'open door' attitudes and policies of that era. Subsequently, migration restrictions have been introduced and the coveted US Green Card has become harder to gain. Nevertheless, around 50 million people live in the USA who were not born there. Over 200 million more are descendants of migrants.

For the USA today, the issue of illegal migration across the Mexican border is a major policy issue that divides the public and politicians alike. While in office, President Barack Obama called for work permits to be issued to many of the estimated 8 million unauthorized workers living in the USA. In contrast, some of his opponents demanded that a wall be built along the Mexican border in order to stop illegal Central American migrants from heading north in large numbers.

The spatial distribution of unauthorized workers in the USA is highly uneven (Figure 5.31). One reason why US citizens often have different views on migration is clearly because some places experience its effects more than others. Migration impacts often become amplified at the local scale and have an even greater effect on places than national-scale data suggest.

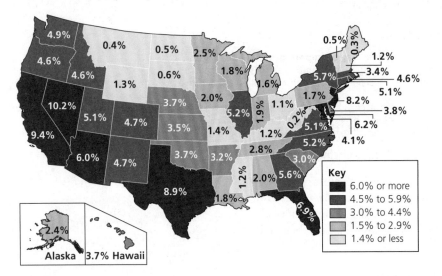

Figure 5.31 Illegal migrants in the USA as a share of the workforce in US states, 2012

The migration issues that commonly divide US public opinion are shown in Table 5.10.

Table 5.10 Reasons for differing views about migration held by US citizens and organizations

Economic impacts	One view is that migrants are a vital part of the US economy's growth engine. From New York restaurant kitchens to California's vineyards, legal and illegal migrants work long hours for low pay. However, high unemployment in some cities has led to calls for American jobs to be given to American citizens instead.
National security	The terrorist attacks on the USA in 2001 ushered in an era of heightened security concerns. Support grew for the anti-immigration 'Tea Party' movement. In 2016, before he became US President, Donald Trump suggested that Muslims should be banned from entering the USA on the basis that global terror group Daesh (IS) pledges allegiance to Islam. Many people found this deeply offensive. Trump insisted he was simply thinking of ways to safeguard national security.
Demographic impacts	In the USA and other developed countries, youthful migration helps offset the costs of an ageing population. Yet the higher birth rates of some immigrant communities are changing the ethnic population composition of the USA. In 1950, 3 million US citizens were Hispanic. Today, the figure has reached 60 million. This is more than one-fifth of the population.
Cultural change	Migrants alter places when they influence food, music and language. Hispanic population growth is affecting the content of US media as programmers and advertisers seek out a larger share of the audience by offering Spanish-language soap operas on channels like Netflix.

Geopolitical constraints on global interactions

Revised

Governments may try to prevent or control global flows of people, goods and information, with varying success.

- Trade protectionism is still common, despite the efforts of the Bretton Woods institutions (Unit 4.1).

- Around 40 world governments limit their citizens' freedom to access online information. Violent or sexual imagery is censored in many countries. However, a 'dark web' also exists, which is harder to control.
- Laws can be strengthened to limit numbers of migrants or to attempt to control how they integrate into society (Figure 5.32). Migration issues are often hard to tackle, though, as the USA and EU have discovered.

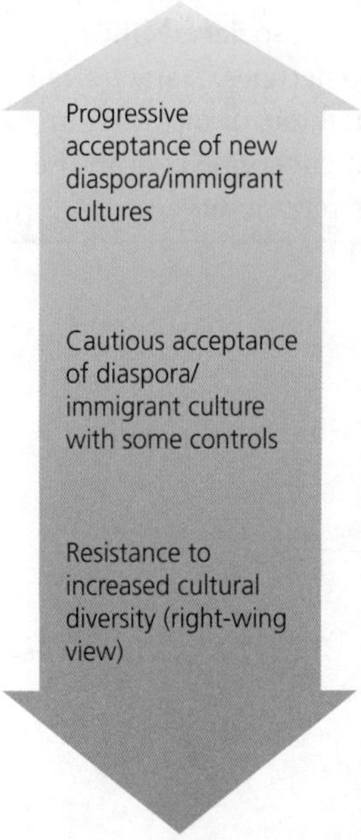

↑ Progressive acceptance of new diaspora/immigrant cultures Cautious acceptance of diaspora/immigrant culture with some controls Resistance to increased cultural diversity (right-wing view) ↓	'Melting pot' (or hybridism)	Positive view of American culture as organic or hybrid — it adopts and absorbs new migrant values
	Pluralism	EU nations tolerate equal rights for all migrants to practise their religious and cultural beliefs
	'Citizenship' testing	UK rules for migrants are becoming stricter in reaction to popular concerns over immigration
	Assimilation	A belief that minority traits should disappear as immigrants adopt host values
	Internet censorship	Preventing citizens from learning about other global viewpoints using online sources, e.g. China
	Religious intolerance	Notably lower levels of religious freedom for minority groups exist in some places, e.g. Iran
	Closed door to migration	Stopping any immigration altogether for fears of cultural dilution, e.g. Cambodia (the Pol Pot years)

Figure 5.32 The cultural continuum: differing government responses to migration and cultural diversity

Government and militia controls on personal freedoms

Non-economic dimensions of political controls on global interactions include restrictions on the cultural and information exchanges associated with the shrinking world (Unit 4.2) and social networking. Worldwide, many governments have introduced restrictions on personal freedoms to participate in global interactions. Figure 5.33 shows two examples of restrictions on internet use operating at either the national or personal scale. Worldwide, around 40 national governments have one or both types of limit in place. Restrictions tend to be greater in **autocratic states** such as the People's Republic of China.

> **Keyword definition**
> **Autocratic state** A non-democratic country where political power is concentrated in the hands of one or more people who may be unelected.

Disconnected states
- Some states limit their citizens' access to cross-border flows of information, resulting in a so-called 'splinternet'.
- Facebook, Twitter and YouTube remain unavailable to Chinese users as part of the 'great firewall of China' (in a parallel example of cultural isolationism, only 34 foreign films are allowed to be screened at Chinese cinemas each year).
- Yet, while there is little external connectivity, 550 million Chinese citizens freely interact with one another within a cyberspace 'walled garden' using local blog sites, such as Youku.
- Other states with similar restrictions include Iran and, increasingly, Pakistan.

Disconnected citizens
- In some states, people additionally lack the means to communicate digitally with their fellow citizens within national boundaries.
- While cost is of course a factor, 25 million North Koreans largely have no access to the internet at all as a result of political decision making.
- In the past, the authorities in Saudi Arabia have restricted messaging using BlackBerry because security forces could not crack the BlackBerry encryption code and were therefore unable to eavesdrop on private conversations. One source for this paranoia was the use of BlackBerrys by the 2008 Mumbai terror attackers, which led to calls for a ban in India, too.

Figure 5.33 Two types of digital exclusion experienced by people and places

The rise of militia groups

According to United Nations data, conflict and persecution have driven more people from their homes in the last decade than at any time since records began. On average, 24 people were forced to flee their homes each minute in 2015, four times more than a decade earlier, when six people fled every 60 seconds. Forced displacements worldwide add up to more than 65 million people. Wars have broken out since 2010 in Syria, South Sudan, Yemen, Burundi, Ukraine and Central African Republic. Thousands more people have fled violence in Central America.

The **militia** groups that play a role in many of these troubled regions have disconnected large numbers of people from global networks by preventing them from interacting with other people and places. This has been achieved in two ways.

1 Some militias deliberately rob civilians of their freedom. Reports of kidnapping, torture and other human rights abuses have become depressingly commonplace in conflict zones.
2 Large numbers of people fleeing militia groups become internally displaced persons (IDPs). Forced to flee their home and possessions, they remain within their country's borders. This can leave people completely disconnected both economically and socially. Children cease to be schooled; adults are unable to work. In 2015, the countries with the largest numbers of internally displaced persons were Colombia (6.9 million), Syria (6.6 million) and Iraq (4.4 million).

> **Keyword definition**
> **Militia** An armed non-official or informal military force raised by members of civil society. Militia groups are sometimes characterized as freedom fighters or terrorists in varying political contexts or in the views of different observers.

Between the mid-1990s and 2010, millions of people fled their homes in DR Congo (see Unit 4.3, page 37) due to conflict and attacks by armed militia groups, including the notorious Lord's Resistance Army (LRA) led by Joseph Kony. During these years, militia groups and crime networks flourished across central Africa where used Kalashnikov assault rifles could be purchased for as little as US$10. The LRA and other groups forced tens of thousands of children to become soldiers. Some as young as ten were forced to murder adults and many suffered trauma as a result. Combined with interrupted schooling, they have often struggled to find employment even after being released from captivity; few international investors have shown much interest in DR Congo due to its human development problems, creating a vicious circle.

While the situation in DRC has improved in recent years, new troubles have broken out in northeast Nigeria, where the Boko Haram militia group's campaign of violence and kidnapping has displaced more than 2 million people.

These themes are discussed further in Unit 6.1, which deals with the phenomenon of 'tribalization'.

■ National trade and investment restrictions

In Unit 4.2 (page 14), we saw that 2016 was the fifth consecutive year when global trade did not grow (as a percentage of GDP). A global-scale slowdown has taken place in some cross-border movements. There are two reasons for this:

- In part, the issue is caused by problems *with the system itself* (Figure 5.34). The world economy has experienced 'boom and bust' phases before. Some economists view the period since 2008 as a new cyclical or permanent downturn; it follows a boom that began in the late 1980s.
- Additionally, new barriers are being raised in some localities to halt global flows *that would otherwise operate*. Some economists fear a return to the 1930s when **protectionism** damaged world trade and contributed to the Great Depression. Populist movements against free trade have strengthened recently in many countries.

> **Keyword definition**
> **Protectionism** When state governments erect barriers to foreign trade and investment such as import taxes. The aim is to protect their own industries from competition.

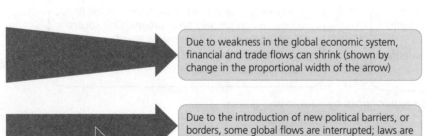

Figure 5.34 Reasons for the changing size of selected global trade and investment flows

Due to weakness in the global economic system, financial and trade flows can shrink (shown by change in the proportional width of the arrow)

Due to the introduction of new political barriers, or borders, some global flows are interrupted; laws are passed which halt trade, migration or data flows

Table 5.11 shows evidence gathered in 2016 that supports the case for rising protectionism on a global scale.

Table 5.11 Recent signs of rising protectionism

Fading enthusiasm for new free trade agreements	A proposed vast new Pacific Rim pact known as the Trans-Pacific Partnership (TPP) has made only very slow progress since it was first suggested in 2006. It involves 12 countries: the USA, Japan, Malaysia, Vietnam, Singapore, Brunei, Australia, New Zealand, Canada, Mexico, Chile and Peru. It has been agreed but not yet ratified – and political support may now be waning (especially in the USA).
Western calls for limits on Chinese investment	The USA raised tariffs on Japanese electronics imports in the 1980s. Now, the US government has taken issue with the way the 'great firewall 'of China blocks access to brands like Facebook and Reuters. They say that US media companies cannot compete fairly for access to Chinese markets. The US has already raised tariffs on Chinese steel imports. European countries have also blocked some proposed Chinese acquisitions while calling for higher tariffs on Chinese imports of, among other things, solar panels (Figure 5.35).
Investment rules for Canada	Companies seeking to acquire a Canadian business must now ask for government approval under the Investment Canada Act. This involves passing a national security test and demonstrating the proposals will have a net benefit for Canada. Some deals have been blocked.
Australia's growing resistance to foreign investment	Australia has tightened rules on foreign property and land purchases. The government has blocked bids for farmland by Chinese investors. It has also moved to block the sale of a controlling stake in the country's largest electricity network, Ausgrid, to a Chinese TNC.

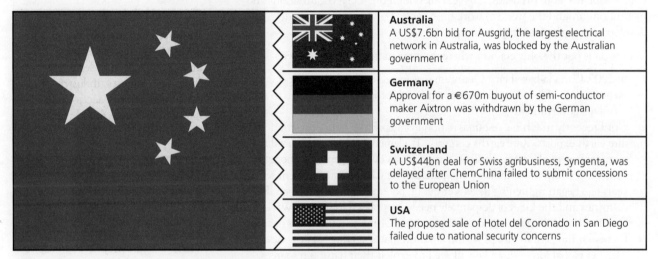

Figure 5.35 Selected barriers to Chinese foreign investment in 2016

■ Other reasons for trade restrictions

Some states have always practised trade restrictions and prohibitions in particular economic sectors. In addition to protecting their own companies, reasons have included:

- *ecological risks* For 50 years, imports of Australian honey were banned in New Zealand for fears of a bio-security threat: Australia's bees suffer from a disease New Zealand beekeepers wanted to avoid. The worldwide export of British beef was banned during the 1990s amid fears over the threat of bovine spongiform encephalopathy (BSE) or 'mad cow disease'.
- *geopolitics* Between 1962 and 2015, the USA imposed a complete trade embargo on Cuba as a result of Cold War antagonism between the two countries. A trade war developed between Russia and the EU after Russia's annexation of Crimea in 2014. Among many other measures, EU oil companies were not allowed to make new investments in Russia, while Russia stopped exporting apples to Poland.

Some developing countries have for many years chosen to restrict their level of trade with developed countries owing to their leaders' scepticism about the fairness of global trade. In January 2017, the charity Oxfam reported that the richest 8 people on the planet now own as much wealth as the poorest 3.6 billion. Facts like this serve to emphasize the point that the global economic system is far from perfect in terms of how it distributes and redistributes the profits resulting from economic interactions between people and places.

Leaders of several South American, African and Asian nations have criticized repeatedly the ways in which some states and people gain disproportionately from global trade and use their power to reproduce this system to their own advantage. Bolivia's Evo Morales, Zimbabwe's Robert Mugabe and North Korea's Kim Jong-un have all spoken out against what they see as global capitalism's rampant inequality. Latin American states including Cuba, Bolivia and Venezuela have even attempted to follow an alternative development pathway. Rather than adopt the free-market capitalism favoured by the USA and EU, these countries have followed socialist principles where possible.

Venezuela's leader Hugo Chávez called repeatedly for other Latin American and Caribbean countries to resist the 'free-market neoliberal order' promoted by the USA and the Washington-based World Bank and IMF. Chávez also used **resource nationalism** as an economic and political strategy. This involved seizing control of oil-producing operations in Venezuela owned by foreign TNCs including Chevron, BP, ExxonMobil and ConocoPhillips. Chávez took the view that Venezuela – which has larger proven fossil fuel reserves than any other OPEC member state – should not give Western TNCs access to its profitable oil. Instead, Venezuela would keep this petrodollar wealth for itself in order to fund education, health and employment not just domestically but in other socialist countries including Cuba. Shortly before his death, Chávez called for more social organizations and institutions around the world to work together 'to build alternative models of development in the face in globalization'.

Venezuela is not the only country where resource nationalism has grown:

- In 2009, Canada-based First Quantum was forced to hand over 65 per cent ownership of a US$550 million copper mining project in DR Congo to the country's government.
- Until recently in China, resource nationalism took the form of restrictions on rare earth exports. Rare earths comprise 17 elements, which are crucial to the manufacture of electronic devices, ranging from everyday mobile phones to strategically important military equipment. The Chinese government several years ago began tightening its export quota, potentially starving Japan, EU countries and the USA of desperately needed supplies. However, a World Trade Organization ruling later succeeded in getting China to reverse this decision.
- Particular indigenous groups within a nation may sometimes take a view on whether global forces should be allowed to exploit their natural resources. Opposition can be strong when an important landscape is threatened by the resource extraction process (see Unit 5.2, page 47).

> **Keyword definition**
> **Resource nationalism** When state governments restrict exports to other countries in order to give their own domestic industries and consumers priority access to the national resources found within their borders.

> **PPPPSS CONCEPTS**
> Think about the alternative possibilities that may exist to the current model of global capitalism. Could globalization evolve in the future to embrace socialist principles instead of neoliberalism and free markets?

Figure 5.36 Venezuela has larger proven fossil fuel reserves than any other OPEC member state

> **Keyword definition**
> **International-mindedness** A way of thinking that is receptive to ideas from different countries and recognizes that all people belong to a networked international community that is pluralistic, culturally diverse and meritocratic. It also involves an appreciation of the complexity of our world and our interactions with one another.

Civil society campaigning

Civil society sometimes plays an important role promoting **international-mindedness** by campaigning for internet freedom. Table 5.12 shows examples of campaigning individuals, places and organizations acting in internationally minded ways. In addition to the case studies provided here, Unit 6.3 looks in detail at the work of Greenpeace and Amnesty International.

5.3 The power of places to resist or accept change

Table 5.12 Examples of civil society campaigning

Individuals and places	Non-governmental organizations (NGOs) and pressure groups
Consumers can refuse to give custom to high-profile high-street global brands like McDonald's, or may boycott polluting air travel.	Amnesty International has opposed the negative impacts TNCs have had on the human rights of some local communities.
Critics of globalization include powerful writers (Naomi Klein, Susan George) and publications (*Le Monde Diplomatique*).	World Social Forum is an annual meeting in Brazil where committed individuals and NGOs campaign for global social justice.
Seattle City Council has a Climate Action Plan that aims to make the city carbon neutral by 2050.	The Electronic Frontier Foundation has been defending free speech online in many different countries since 1990.

CASE STUDY

CHAMPIONING RESTRICTED INTERNET FREEDOMS

In Unit 5.1 (page 46) we learned how affirmative action in support of women's education had been taken in Pakistan's Swat Valley. Part of championing women's participation in education involves freedom to participate in global interactions online. You will know from your own experiences how essential the internet has become as an educational tool. Students across the world rely increasingly on formal global news sites like CNN and Al-Jazeera in addition to Wikipedia's 'collective commons' information store. As we have seen, however, some states do not grant citizens unfettered access to the internet, preferring instead to develop a 'splinternet', which is only partially integrated into the worldwide web. Table 5.13 shows two examples of places where this happens and actions that have challenged these restricted freedoms. The data were provided by the OpenNet Initiative (a collaborative project involving, among others, Harvard University).

Table 5.13 Challenging restricted internet freedoms in Iran and China

	Internet restrictions	Civil society challenges
Iran	• Citizens in Iran face severe restrictions on internet connectivity according to OpenNet. Internet searches are filtered (meaning access to some information is not allowed) and it is not always possible to use YouTube or social networking sites like Twitter. • Recently, Iran's government has been taking steps towards the introduction of a National Information Network (NIN) which would 1) increase its control over how citizens can use the internet and 2) restrict foreign users from accessing Iranian websites. • The Iranian government has also taken steps that make it much harder for citizens to use the internet anonymously.	• Internet freedom in Iran is championed by the International Campaign for Human Rights in Iran. Founded in 2008, this organization is based in New York and works with Iranian political activists. They hope to persuade the Iranian government to soften its attitude towards the internet. • Iranian political activists have used the internet to challenge its government before (one reason why the NIN is now being introduced). When a young woman called Nedā Āghā-Soltān was shot dead during the 2009 Iranian election protests, footage of the incident went viral on YouTube and Facebook. It helped flame protests that lasted for ten days.
China	• The so-called 'great firewall' of China consists of extensive controls on internet use, which have created a 'walled garden' for citizens: people inside China can interact with one another but cannot use global social networking sites. Instead of Facebook, hundreds of millions of Chinese internet users interact using Qzone. • In 2010, Google withdrew from China temporarily due to censorship issues; today, most Chinese internet users use the domestic service Baidu. Chinese internet users will not find information about the 1988 Tiananmen Square massacre on Baidu: the government has censored all references to it. • Foreign news sites like the BBC and CNN are strictly monitored and certain news items removed.	• GreatFire is a Chinese citizen-led organization dedicated to fighting internet censorship in China. GreatFire says it gets its funding from people and organizations inside China and also the Open Technology Fund (a US-government-backed initiative to support internet freedom). • Recently, the US government has begun arguing that China's internet controls amount to a trade barrier for e-commerce. Data flows and services have become such an important part of globalization that internet barriers may increasingly be seen to breach World Trade Organization rules. This is an ongoing source of dispute between the two superpowers. • China looks likely to face escalating criticism from other countries that the great firewall is an impediment to free trade as much as to free speech.

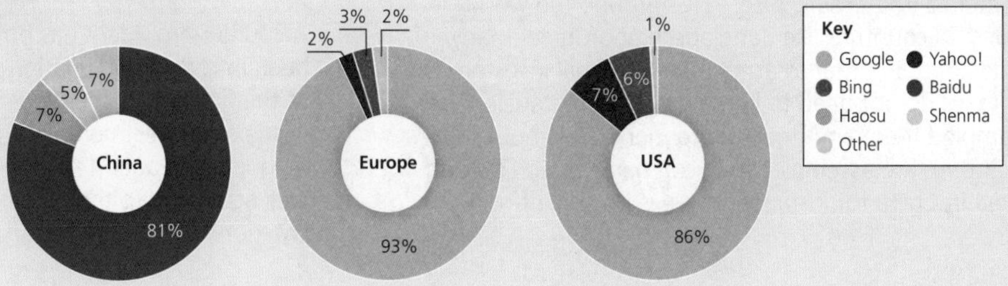

Figure 5.37 Variations in search engine market share for different countries and regions, 2016

China: 81%, 7%, 5%, 7%
Europe: 93%, 3%, 2%, 2%
USA: 86%, 7%, 6%, 1%

Key: Google, Yahoo!, Bing, Baidu, Haosu, Shenma, Other

Unit 5 Human development and diversity

■ KNOWLEDGE CHECKLIST

- The concepts of civil society resistance and nationalism
- Examples of campaigns against TNCs
- The local sourcing movement for food and goods led by citizens
- The rise of anti-immigration movements
- Contemporary geopolitical constraints on global interactions
- Government controls on personal freedoms to participate in global interactions
- Militias and the way they curtail people's freedoms
- Reasons for national trade restrictions
- Examples of protectionism and tariffs
- The concept of resource nationalism and examples of when it has occurred
- The concept of international-mindedness and global participation
- Social media use and campaigning for internet freedom for citizens, using examples of places where restricted freedoms have been challenged
  (China, Iran)

EVALUATION, SYNTHESIS AND SKILLS (ESK) SUMMARY

- How acceptance of, or resistance to, global interactions, takes different forms
- How acceptance of, or resistance to, different kinds of global flow occurs at different geographic scales

EXAM FOCUS

SYNTHESIS AND EVALUATION WRITING SKILLS

Below is another sample HL answer to a part (b) exam-style question. You will recall that this type of question requires you to **synthesize** and **evaluate** your information (under assessment criterion AO3). As you progress through the course, a greater volume of ideas, concepts, case studies and theories become available for you to synthesize. This sample answer is drawing on ideas from all of the first two units of this book. The levels-based mark scheme is on page vii.

The question has a maximum score of 16 marks. Read it and the comments around it.

Barriers to globalization are on the rise in many parts of the world. Discuss this statement. (16 marks)

Globalization is the lengthening and deepening of connections between people and places on a global scale. There are economic, social, cultural and political dimensions to this process. It is important to recognize this because while barriers may be rising against certain aspects of globalization in some places, it may not be the case that all aspects of globalization are being rejected. (1)

1 This is a clear introduction that deconstructs the key term and establishes a discursive structure.

Looking firstly at economic globalization, there is plenty of evidence to suggest that barriers to global trade flows have recently increased in some places. This is called protectionism. Back in the 1930s this was a major cause of the Great Depression. Since the Global Financial Crisis of 2008, slower economic growth has led some governments to try and protect their industries in order to save jobs. (2) Many European countries and the USA have been calling for greater import taxes on Chinese steel. When the Port Talbot steel works in Wales, UK were threatened with closure in 2016, 30,000 jobs were put at risk. (3) It is easy to see why governments might be thinking twice about free trade.

2 There is good recall of terminology here.

Progress has also been slow in finalizing some large-scale trade agreements such as the EU's new deal with Canada which took seven years to complete due to objections from Belgium.

3 The ideas are supported by the use of detailed examples.

Despite this, free trade is still taking place in many large trading areas like the European Union and South America (Mercosur). Also, TNCs around the world continue to outsource and invest in countries like Bangladesh and China. Many Asian countries have free-trade zones in order to attract international investors. These are sites where import taxes do not need to be paid on raw materials and parts. Goods can be assembled there and exported to markets without encountering trade barriers. iPhones are assembled by a company called Foxconn in the Shenzhen free-trade area in China for instance. There is no sign of rising barriers here. (4)

4 A counter-argument has now been introduced in relation to the first global theme (trade and investment). This is 'ongoing evaluation' and scores well according to the AO3 criterion.

Turning next to migration, the evidence is also unclear as to whether barriers are rising or falling – perspectives differ on this. (5) Recently the UK voted to leave the European Union on account of many British citizens' desire to reduce migration flows into the country. Many other European countries, such as France, have seen growing support for anti-immigration movements too. In the USA, many citizens are opposed to migration from Mexico and the rest of Latin America. In his presidential campaign, Donald Trump vowed to build a wall along the Mexican border to stop migration. (6)

However, in other parts of the world there is growing enthusiasm for free movement of people and visa-free travel. The African Union is taking steps to make movement easier for all 54 of its member states. South American countries have also agreed to make temporary residency rights easier to gain. Therefore it is unclear overall whether global migration is on the rise or decline. A record number of 250 million people currently live outside the country they were born in and it seems unlikely that this number will fall any time soon. (7)

Finally, let us look at global data and information flows. These play a role in economic globalization because services and goods are increasingly traded internationally online. Data flows also contribute to cultural globalization by sharing music, ideas, languages and other aspects of culture. It is true that internet freedoms are threatened in many places. Iran may soon join China in having its own 'splinternet'. This is an area of the internet that is walled off from the rest of the world. Despite this total global data flows are at an all-time high. Smartphones and Facebook are barely a decade old. (8) Their use increases every year as more people are able to afford phones and more people participate in social networks which are global in scale. I have over 200 Facebook friends and they include people in Singapore, New Zealand, Australia, the Netherlands, the USA, Canada and Georgia. This makes me feel extremely globalized. Many other Millennials like myself regard ourselves as being global citizens. It may be there is a demographic divide between older people with nationalist attitudes who want to retreat from globalization, and my generation, more of whom support it. (9)

In conclusion, the extent to which barriers are rising depends on what aspects of globalization you are looking at, the place contexts you study, and the age and perspectives of the people involved.

5 A new theme is introduced to the discussion here, with opposing perspectives acknowledged from the outset.

6 This paragraph makes good use of contemporary events and case studies.

7 Once again, there is ongoing evaluation, grounded in evidence. Strong counter-arguments have been made about migration.

8 A third theme is introduced; again, the material is well-argued, evaluative and well-evidenced.

9 Here, the writer offers his or her own perspective. Personal experience and personal geographies can always be drawn on in an essay, and this point is clearly relevant to the discussion.

Examiner's comment

The quality of synthesis and evaluation (AO3) here is excellent. The conclusion is rather short; however, there has been plenty of ongoing evaluation throughout the essay, which compensates for this. Overall, there is no reason why this answer should not reach the highest mark band.

Content mapping

You are now two-thirds of the way through the course and there is a large amount of material that can be drawn on to help answer this essay. This answer has made use of case studies, concepts, theories and issues drawn from all of the first two units of the book. Using the book's index, try to map the content.

Unit 6 Global risks and resilience

6.1 Geopolitical and economic risks

You may be very familiar already with the concept of **risk** from studying geophysical or hydrological hazards elsewhere in geography. In the context of the study of global interactions, physical threats are just one among several risk categories. Figure 6.1 shows five categories of physical and human risks that states, businesses and societies are increasingly exposed to.

> **Keyword definition**
> **Risk** A real or perceived threat against any aspect of social or economic life.

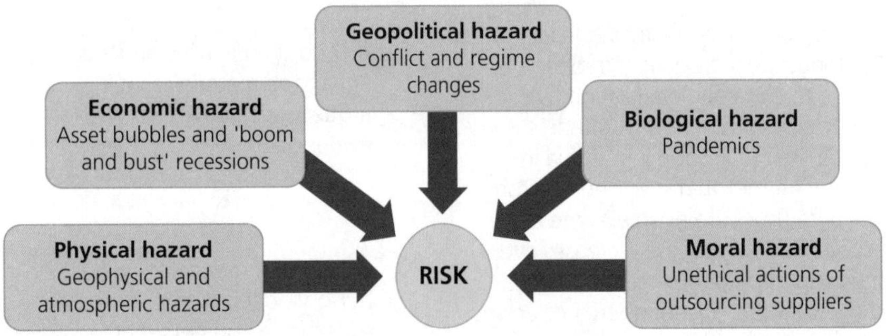

Figure 6.1 Globalization has meant that states, societies and businesses face an increasing number and range of hazard risks

- **Geopolitical hazard**: Conflict and regime changes
- **Economic hazard**: Asset bubbles and 'boom and bust' recessions
- **Biological hazard**: Pandemics
- **Physical hazard**: Geophysical and atmospheric hazards
- **Moral hazard**: Unethical actions of outsourcing suppliers
- **RISK**

Some human and physical hazards and risks may be avoided through mitigation measures; others may be unpreventable and unavoidable. Worst of all are 'black swan' events: these are 'unthinkable' high-impact, hard-to-predict and rare occurrences. In an interconnected world, black swan events – such as the earthquake and tsunami which struck Japan in 2011 or the Global Financial Crisis (GFC) of 2007–09 – bring disproportionate impacts. They can make us question the wisdom of globalization on account of the sheer number of negative 'knock-on' effects for different places and societies.

The global community's growing vulnerability to risk is a function of increasing interdependency: events in one part of the world now have effects on distant places and people.

New threats created by technological and globalizing processes

Multiple hazard risks accompany the new economic opportunities and social freedoms brought by internet and social network growth. Individuals who make regular use of these technologies are also exposing themselves to numerous threats, some of which you may have personal experience of:

- Text messages, email and online comments are sometimes used wrongly to harass people. School children may fall victim to cyber-bullying. Many celebrities, politicians, journalists and political campaigners have suffered from so-called 'trolling', which involves hateful comments being posted online.
- Spam email attachments may sometimes contain harmful computer viruses.
- Widely available pornographic and violent online imagery can do psychological harm to vulnerable viewers.
- Extremist groups and militias such as Daesh recruit new members online.
- Young bloggers sometimes encourage one another to self-harm or engage in other dangerous activities such as excessive weight loss.

It is not just individuals who have sometimes suffered harm on account of information and communications technology (ICT) use. The roll-out of new digital technology has undermined entire industries: think of how the music and film industries have suffered great profit losses on account of free file-sharing in a 'post-DVD' world, or the unbeatable price competition Amazon has brought to high-street retailers.

■ Hacking, identity theft and personal freedom

Dealing with cyber-crime is an increasing priority for businesses and law enforcement agencies globally (Figure 6.2). The speed with which new technology develops means that the boundaries keep changing, however. No sooner have defences against the latest viruses and spyware been introduced than newer threats emerge to take their place! Table 6.1 shows some recent examples of hacking and identity theft.

Table 6.1 Examples of hacking, identity theft and security breaches

WikiLeaks	Julian Assange's WikiLeaks is a well-known organization linked with computer hacking. In the 1990s, Assange is alleged to have hacked into various Australian and US defence department's computer systems, including MILNET, the US military's secret defence data network. More recently, WikiLeaks has courted controversy by collecting and publishing a range of secret and classified government information.
Yahoo	Media TNC Yahoo suffered the biggest data security breach in history in 2014. Details for 500 million Yahoo accounts were stolen, including people's names, email addresses, phone numbers, birth dates and, in some cases, security questions and answers. In a press release, Yahoo said it believed the hackers had been 'state-sponsored' meaning that a state government may have been involved.
Tesco	The UK retail and banking TNC Tesco suffered an unprecedented security breach in 2016. Cyber-attackers breached the company's bank's central system and stole money from 20,000 bank accounts.

- 25% of all hacking attacks come from within the USA
- 16,000 websites hacked each year in india
- 15% of hacking attacks originate from inside China
- over 10% of hacking attacks come from Russia
- 40% of the world's population has internet access; in 1995, it was less than 1%
- It is estimated that more than 50 million computers are infected by a virus
- 90% of businesses have been hacked at some point
- Computer hackers steal more than US$1 billion annually from banks

Figure 6.2 A global view of hacking (2012 data estimates)

■ Personal freedom in a surveillance society

Parallel to the risk of hacking and identity threats, ICT also threatens personal freedoms in what is sometimes described as 'the rise of the surveillance society'. Sophisticated software and communications can be used by the state to monitor the activity and behaviour of citizens. One view is that countries are at risk increasingly of becoming what George Orwell called 'big brother' societies (Figure 6.3). The view of civil liberties supporters is that just because we can use technology increasingly to monitor daily life it does not mean we *should*.

On the one hand, there is often a perfectly sound rationale for introducing CCTV in real space and monitoring what happens in virtual space more closely:

- CCTV cameras can help to reduce crime and improve safety for vulnerable people, particularly at night.
- Speed cameras make sure motorists do not pass safety limits; in the near future, cars may be controlled and moved remotely by centralized computers in some cities.
- National security issues can help to make a case for greater online surveillance. Unit 5.3 (page 75) explored reasons why some states have placed restrictions on how citizens can use the internet: in some cases, quite reasonable-sounding arguments about national security and citizen safety have been put forward to justify online controls and censorship.
- There may be popular support for governments being allowed to 'snoop' online. A worldwide moral panic has arisen over concern that deviant social groups – from paedophiles to terrorists – are planning or participating in criminal activity online, far from the prying eyes of law enforcers.

Figure 6.3 Are we living in 'big brother' societies?

Global supply chain risks

TNCs operating in every sector of industry can find that the benefits gained from expanding globally are countered by the costs of interrupted trade: the more extensive a company's network becomes, the greater the number of possible local threats it becomes exposed to (Table 6.2).

Table 6.2 Recent examples of global risk exposure for TNCs

Geopolitical and conflict risks	Political events and changes in foreign government policies can create supply chain shocks for TNCs: • In 2014, Foxconn was forced to suspend production of iPhone components in Vietnam owing to political demonstrations taking place there. Yue Yuen, the world's biggest sports-shoe maker, which supplies Nike and Adidas, shut down temporarily. • US and EU oil companies operating in Russia were told by their governments to suspend any new operations there after sanctions were imposed following Russia's annexation of Crimea operations in 2014. • Many tourist and airline companies have lost custom as a result of conflicts and political instability: tourist revenues in Tunisia halved between 2010 and 2014 on account of Daesh terror attacks targeting holidaymakers. • The 'Arab Spring' wave of uprisings in North Africa in 2011 meant French companies such as France Télécom experienced service and supply chain disruption across French-speaking North Africa (work stopped at the Teleperformance call centre in Tunis, for instance). • Resource nationalism can result in a TNC's overseas operations being seized by a state government (see Unit 5.3, page 78).
Moral and ethical hazard risks	Unethical treatment of supply chain workers and environments by outsourcing companies jeopardizes the reputation of the TNCs who do business with them: • In highly publicized 2010 court cases, European companies BP and Trafigura both tried to lay the blame for catastrophic environmental damage (in the Gulf of Mexico and Ivory Coast, respectively) at the doors of subcontractors. Ignorance of harm done by a subcontractor to people or the environment rapidly translates into a 'moral' and reputational hazard situation for the hub company. • Reports of child labour or worker suicides (China, 2010) can do enormous damage to a company's reputation. A spate of fires in Dhaka during 2010 exposed clothing companies Gap and H&M to brand association with poorly monitored and unsafe factories owned by the subcontracting firm Hameem (in the worst case, 26 workers died when fire exits were blocked). • As we saw in Unit 5.3, the Dodd–Frank Act resulted in many US TNCs removing DRC from their supply chain altogether rather than risk being associated with conflict minerals.
Physical risks	Natural hazards can disrupt supply chains unexpectedly: • Millions of cubic metres of volcanic material known as tephra were ejected over Iceland when Eyjafjallajökull erupted in 2010 (Figure 6.4). Fights were grounded for weeks after vast amounts of fine ash rose 9 kilometres into the paths of jet aeroplanes. Kenyan farmers and flower suppliers went out of business because they couldn't get products to market in the UK (5,000 supply chain workers were temporarily laid off). Car production in Europe also ground to a halt, as the flight ban starved firms like Nissan of key parts. • Thailand, a hub for electronics and car parts, suffered its worst floods in 50 years in 2011. Computer makers struggled to get hard drives while automakers such as Honda were forced to cut global production.

Figure 6.4 The Eyjafjallajökull eruption of 2010

CASE STUDY

SUPPLY CHAIN DISRUPTION: JAPAN'S 2011 TSUNAMI

Japan's magnitude 8.9 earthquake and tsunami highlighted the overexposure of firms across the world to environmental supply chain risks. It was also the cause of a major transboundary pollution event on account of a nuclear power station meltdown (see Unit 6.2, page 95). The devastating disruption caused by the earthquake and tsunami hit company profits hard both within and outside the country.

- Worst hit were Japanese manufacturers such as the carmakers Honda, Toyota and Nissan, as well as technology groups like Sony, Nintendo and Panasonic. Supply chain disruption for these domestic firms was severe.

- Some foreign-owned manufacturing companies suffered significant production setbacks. US carmakers Ford and Chrysler slowed production of red and black vehicles in 2011 – because the factory that was the sole supplier of a paint pigment vital for the production of these colours was in the tsunami zone.

- The earthquake brought major disruption to global supply chains in high-technology industries such as flat-panel TV manufacturing. Japan is the world leader in precision machinery parts (making 100 per cent of all protective polarizer film for LCDs and around half of the world's small car motors).

- Japanese technology companies account for 40 per cent of the world's information technology component supply, including 30 per cent of flash memory for smartphones and cameras. US firm Hewlett-Packard lost US$700m revenues due to supply chain disruption and falling demand in Japan.

- Mitsubishi Gas Chemical (MGC) ceased production of bismaleimide triazine resin (a key material used for smartphones and tablet production; Japan supplies 90 per cent of the world's specialist resins used in the semiconductor industry). Falling global production of smartphones was widely reported (Figure 6.5).

- Korean Air Lines suffered operating losses when flights in and out of Japan were cancelled. But the South Korean carmaker Hyundai saw a 37 per cent rise in sales as it took advantage of the disrupted production of its Japanese rivals.

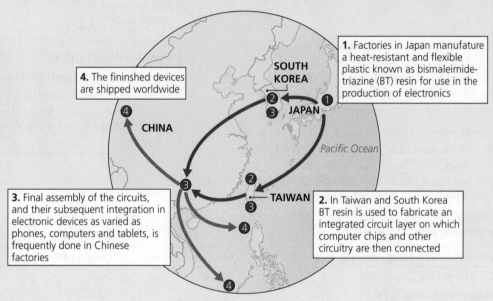

Figure 6.5 Japan's key role in global smartphone production

New threats to state sovereignty

The rationale for retreating from globalization is rooted, for some people, in the valid concern that actual **political sovereignty** and/or **economic sovereignty** has been, or is being, surrendered. It is certainly true that multi-governmental organizations (MGOs) like the EU, NAFTA and Mercosur have gained power over certain aspects of national life, including trade and migration rules (see Units 4.3 and 5.3).

The renewal of nationalism is also linked with a much broader interpretation of 'loss of sovereignty', which is synonymous with 'loss of control'. The shrinking world, so the argument goes, has brought chaos to national life and culture due to new and sometimes unchecked flows of people, information and ideas.

> **Keyword definitions**
>
> **Political sovereignty** The freedom of a state to govern itself fully, independent of interference by any foreign power. In theory, no United Nations member has complete political sovereignty.

War and conflict have always threatened the political sovereignty of states. Since 1945, many states have been attacked or undermined by a rival power, or have been occupied by an invading army or peacekeeping forces.

- The United Nations deployed peacekeepers on 71 occasions between 1948 and 2016.
- The USA has sometimes been described by its critics as a 'rogue superpower' on account of the number of occasions it has made foreign policy decisions that have undermined the political sovereignty of other states (see Figure 6.6). See also Unit 4.1.

> **Keyword definitions**
>
> **Economic sovereignty** The freedom of a state from any outside intervention in its markets and trading relationships. In reality, no state has complete economic sovereignty owing to the complexities of world trade and trading agreements.

Figure 6.6 Selected examples of US intervention in the affairs of other sovereign states since 1945

The acquisition of a country's businesses and infrastructure by foreign buyers could be viewed as a threat to its economic sovereignty. Some state governments buy foreign assets using their **sovereign wealth funds (SWFs)**. These are global-scale 'piggy banks' used by these countries to help build global influence and diversify their income sources.

- Only a minority of countries operate SWFs (Table 6.3). They are mostly countries with oil and gas revenues (such as Norway and Qatar) or mineral resources (Chile's copper wealth and Botswana's diamonds).
- Some governments make purchases directly (Kuwait's US$500 billion fund, dating back to 1953, works this way); others have founded purpose-built investment banks to manage their purchasing (Singapore has two, called Temasek and GIC).

> **Keyword definition**
>
> **Sovereign wealth funds (SWFs)** Money used by state governments to purchase large overseas assets such as power stations and rail infrastructure.

Table 6.3 States with large sovereign wealth funds

Country	Sovereign wealth fund name(s)		Assets (US$ bn)	Origin of wealth
China	1	China Investment Corporation	1,500	Trade
	2	National Social Security Fund		
Abu Dhabi (UAE)	1	Abu Dhabi Investment Authority	1,000	Oil
	2	Abu Dhabi Investment Council		
Norway		Government Pension Fund	900	Oil
Kuwait		Kuwait Investment Authority	600	Oil
Singapore		Government of Singapore Investment Corporation (GIC)	500	Trade
Qatar		Qatar Investment Authority	300	Oil
Australia		Australian Future Fund	100	Oil
Russia	1	Reserve Fund	160	Oil
	2	National Welfare Fund		

As a result of SWF activity, some states have sold certain assets and infrastructure to another state's government. Examples include:

- Singapore owns almost half of New Zealand's Viaduct Quarter (a major residential and commercial development in Auckland).
- Angola is investing its US$5 billion oil-derived assets in infrastructure and hotels in neighbour states.

> **PPPPSS CONCEPTS**
>
> Think carefully about the important difference between asset purchasing by a foreign *TNC* or a foreign *government*: these are quite different processes.

- Norway, which has the single largest SWF, is a major Facebook shareholder.
- China owns numerous assets in the UK, including shares of railways, sewers, airports and financial districts (Figure 6.7).
- The UK's Manchester City Football Club is now owned by Abu Dhabi SWF.

Not all citizens agree that foreign SWF purchasing of their own state's assets should be allowed. Critics may view these sales to foreign state governments as a loss of sovereignty.

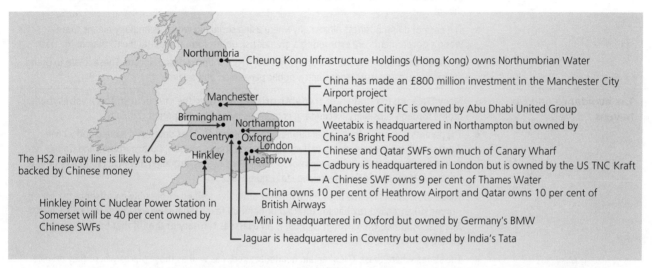

Figure 6.7 Is the UK's sovereignty threatened by the purchase of its assets by other states' sovereign wealth funds?

Profit repatriation and tax avoidance by wealthy TNCs and people

A complex relationship exists between each world state and the TNCs that are domiciled there. Governments and businesses sometimes work together closely in order to further one another's interests. This is another example of interdependency.

- The taxes that TNCs pay in the states where they are domiciled (headquartered) play a vital role in providing governments with much-needed money for health, education, welfare and defence spending. Apple, which is headquartered in California, paid US$6 billion to the US government in 2012; the top taxpayer was ExxonMobil, which handed over US$31 billion. These sums include **profit repatriation** from other states.
- In return, TNCs sometimes look to the national government where they are headquartered for support during a financial crisis, or when their overseas assets become threatened by conflict or nationalization. The oil company Repsol sought support from its country of origin, Spain, when Argentina's government seized control over Repsol's Argentinian investments (in a display of resource nationalism). During the GFC, General Motors and Chrysler looked to the US government for support, while the Royal Bank of Scotland was 'bailed-out' by the UK Treasury.
- Richard Nixon worked as an international lawyer for Pepsi-Cola before becoming US President in 1969. During Nixon's time at the White House, the Soviet Union granted Pepsi-Cola permission to begin manufacturing drinks there. Nixon was a key player in brokering this unprecedented deal in 1972. In 2014, Pepsi-Cola – which has remained headquartered in New York despite the existence of tax havens elsewhere – gave the US government approximately US$2 billion in taxes. Over time, Pepsi-Cola and the US government have both prospered through this interdependent relationship.

What happens, however, if a TNC tries to reduce the tax it pays? State governments are put at risk of losing much-needed income by the two strategies shown in Table 6.4. Some TNCs use these globalizing strategies to reduce their tax burden and increase pay-outs for shareholders instead.

> **Keyword definitions**
>
> **Profit repatriation** A financial flow of profits from a country where a TNC has overseas operations back to the country where its headquarters are.
>
> **Corporate migration** When a TNC changes its corporate identity, relocating its headquarters to a different country.
>
> **Transfer pricing** A financial flow occurring when one division of a TNC based in one country charges a division of the same firm based in another country for the supply of a product or a service. It can lead to less corporation tax being paid.

Table 6.4 Strategies used by some TNCs to limit the amount of corporate tax they pay in the country where they are headquartered

Corporate relocation and profit repatriation	• A TNC may consider leaving its traditional home if the tax regime in another state looks more attractive; this process is called **corporate migration**. In recent years, some European-based TNCs have relocated their headquarters to Ireland, Switzerland, Luxembourg or the Netherlands where corporate taxes are low. In 2010, petrochemical firm Ineos moved its headquarters from the UK to Switzerland. This corporate migration yielded an estimated saving of almost half a billion pounds over five years. Ineos's worldwide profits are now repatriated to Switzerland instead of the UK. • The ease of doing business almost anywhere using shrinking world technology means that the world's governments are exposed to a greater risk of this form of 'capital flight' than in the past. • However, there are practical reasons why many TNCs choose not to relocate. These relate to brand authenticity, corporate responsibility, public perception and security.
Tax avoidance and tax havens	• Many TNCs use the strategy of **transfer pricing** to reduce their tax burden. This involves routing profits through subsidiary (secondary) companies owned by the parent company. • The parent company is the original business that a global TNC has developed around and whose directors still make decisions that affect the organization as a whole. Both Starbucks and Google are parent companies to global networks of subsidiary businesses, including Ritea Ltd (Starbucks Coffee Company, Ireland) and Google Ireland Ltd (Figure 6.8). • These subsidiaries will be based in a low-tax state like Ireland or possibly an offshore tax haven. Around 40 so-called tax havens offer nil or nominal taxes. Some are sovereign states, such as Monaco. Another, the Cayman Islands, is an overseas territory of the UK that has its own tax-setting powers. • The states where these TNCs remain headquartered – as well as those places where they do most business – are at risk of receiving virtually no tax!

It is not just companies who route money in this way. Some wealthy individuals attempt to limit their personal tax liability by migrating to a tax haven where they may either apply for citizenship or become an **expatriate**. Figure 6.9 shows privately owned money flows directed towards low-tax destinations in 2013. Increasingly, international organizations and groups (see Unit 4.1) are taking action to share information about this kind of tax evasion.

> **Keyword definition**
> **Expatriate** Someone who has migrated to live in another state but remains a citizen of the state where they were born.

- The European Union now requires member states to exchange any information they may have.
- More than 80 countries have agreed to share information as part of an OECD-led initiative.

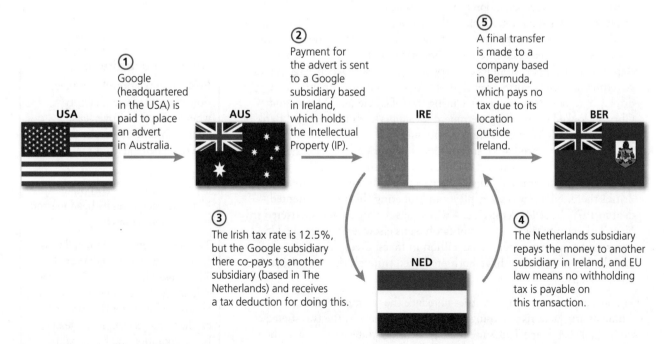

Figure 6.8 A transfer pricing strategy used by some TNCs to reduce their corporate tax bill

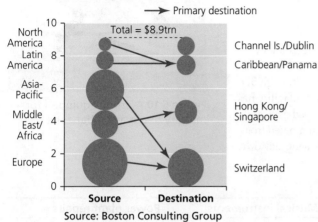
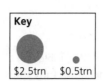

Figure 6.9 Flows of private wealth directed towards popular low-tax regimes and tax havens in 2013

■ Disruptive technologies

New technologies sometimes have unexpected and disruptive effects on societies. Drones and 3D printing are already transforming many aspects of life at both local and international scales. In the near future, we can expect to see artificial intelligence (AI) becoming a third major **disruptive technology** with far-reaching geographic effects.

> **Keyword definition**
> **Disruptive technology** A technology which brings major changes to the way people live and work instead of merely supporting and enhancing the current way things are done.

■ The impact of drones on state sovereignty

Drones are already transforming geographic relations between people and places in a range of contexts (Table 6.5).

Table 6.5 Different drone user groups

User group	Type of use	Evaluation
Individuals	People can mount cameras on their own personal drones and make aerial films	The benefits must be weighed against the threat to other people's privacy
Companies	Amazon expects to use drones to deliver goods to customers before long	Increasing use of drones may threaten the safety of passenger aircraft
Civil society	Drones are being used to help locate victims in earthquake and hurricane zones	Drone technology is already beginning to save lives in this context
Military	Increasingly, conflict can be waged remotely using armed drones	Experts continue to question the legality of drone warfare under international law

Drone technology has become controversial because of the way it is now used in conflict situations, especially by the US military (sometimes assisted by the UK and France).

- The USA's drone missile attacks in Pakistan, Yemen and Somalia – part of its global 'war on terror' – are highly debated.
- The killing of people by unmanned drones raises a challenge for international law makers by arguably breaking Article 51 of the UN Charter (which relates to self-defence).
- In a practice that can be unfavourably compared with computer gaming, 'hellfire missile' operators in Nevada, USA, fly remote-controlled missiles into northern Pakistan.
- The intended targets are suspected terrorists; as many as 3,500 people were reported to have been killed by drones in 2015, around 10 per cent of them civilians.

The rules that govern this kind of conflict are disputed. According to one view, the USA is executing people for crimes they may or may not have committed without a trial and is in breach of the Universal Declaration of Human Rights. However, the USA claims its drone actions are legitimized by war law – despite the fact that the USA is not at war with any of the countries where drone attacks currently take place.

Many people are uncomfortable with the way this latest shrinking world technology is being used and the way in which it quite arguably violates the sovereignty of states and the human rights of people.

■ New geographies of 3D printing

3D printing is revolutionizing many areas of life (Table 6.6). This technology 'paints' layer upon layer of a resin or polymer, until fully three-dimensional objects are created. Around 300 layers of resin create a shape 1 cm thick. Although 3D technology has existed since the 1980s, a recent step-change in affordability has taken place. Since 2010, the price of a basic printer – an oven-shaped device capable of creating objects such as cups or simple statues – has plummeted from around US$10,000 to less than US$1,000. Your school art department may own one already, or will do soon.

Figure 6.10 A drone equipped for military use in a combat zone

Table 6.6 Varied uses of 3D printing

Aircraft manufacturing	Bone replacement	Musical instruments	Power plant repairs
• Fuel nozzles have been printed for Boeing and Airbus aircraft • EADS, Europe's biggest aerospace company, aims to start printing titanium parts several metres in length • Rolls-Royce jet engines will soon be built using some 3D-printed components	• A CAT scanner can perfectly record the dimensions of a person's skeleton • New 'bones' can then be printed using titanium powder with a ceramic coating • In 2012, a Belgian woman received a replacement jaw bone	• A medical CAT scanner was used to reproduce a Stradivarius violin – 1,000 scanned cross sections were used to print an exact replica made of resin • 3D-printed guitars with intricate hollow body designs are already on sale in shops	• Highly specialized replacement parts have been printed to help the repair and decommissioning work at the UK's Sellafield nuclear power station • Titanium and aluminium powders are used, resulting in huge cost savings

3D printing has important implications for state sovereignty and the ability of governments to control what can and cannot pass across state borders. For instance, it may be easier for a state to intercept an illegal shipment of guns than it is to intercept a blueprint for making a gun that has been emailed to a 3D printer (Figure 6.11).

An incredible technological disruption is taking place: physical commodity movements are being replaced by information exchanges, which are often harder to control. Remember too that states raise money through export sales and through taxes levied on imports. If physical trade ceases – to be replaced by information flows directed towards 3D printers – what will happen to those revenue streams? Governments could lose vital revenue from import tariffs. The German electrical company Siemens already envisages delivering spare parts by email to overseas customers, including many living in the USA. It is unclear whether the US government would be able to levy an import tax on 'virtual' spare parts in the way it currently does on real spare parts.

> **PPPPSS CONCEPTS**
>
> Think about how this brief analysis of the geography of technology shows that the balance of power between different people, organizations, governments and places is beginning to change in unpredictable ways.

Figure 6.11 Governments may struggle to control the movement of data that allows 3D guns to be printed

Globalization and tribalization

Unit 5.1 explored the resurgence of **populist** and nationalist movements in different places. Other commentators have established a common link between a wide range of localized movements, including 1) conflict in the Middle East and North Africa and 2) support for anti-immigration parties and policies in the USA, UK, France and other developed countries. In all of these contexts, there is rising skepticism that globalization is a force that remakes places and societies for the better; instead some groups of people view it as a process which has bettered the lives of others but made their own worse.

The result has been what sociologists and psychologists call the **tribalization** of politics. Resistance movements have grown within countries that oppose 'business as usual' politics and support for globalization. Instead, a growing number of citizens appear to be adopting a new defensive form of identity politics.

Figure 6.12 shows evidence that may help explain this trend. The Milanovic 'elephant chart' identifies two groups of people who have fared especially poorly from the past two decades of global growth:

- The poorest people in developing countries such as Chad, DRC and Eritrea, where there has been limited foreign investment (in contrast to the recent success of emerging economies).
- Those citizens of richer developed countries who are most likely to define themselves as 'blue collar', 'working class' or 'ordinary people'.

According to Milanovic's data, neither of these groups has experienced a rise in real incomes since 1988, while others have benefited more. As a result, the argument goes, so-called tribalization politics are strongest in these groups.

> **Keyword definition**
>
> **Tribalization** The rise of 'us and them' political movements, which are often opposed to globalization or Westernization.
>
> **Populism** The idea that every political decision in a democracy should reflect what the majority of citizens believe, not what the majority of politicians believe.

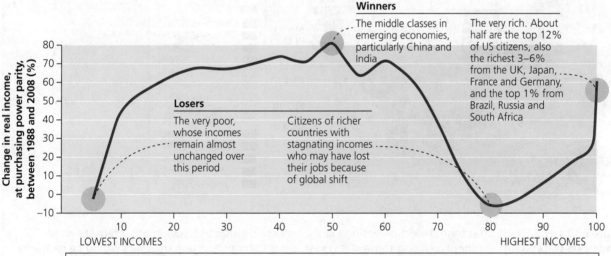

- The x-axis represents the world's 7 billion people arranged in order of their income size
- The y-axis shows how their income grew during the 'golden age' of globalization (1988–2008)
- Based on this, which people and places may we infer could be most – or least – supportive of globalization, and why?

Source: Branko Milanovic

Figure 6.12 The Milanovic 'elephant chart' shows globalization's winners and losers

What next for globalization?

It is important to remember that even those groups who claim to be opposed to globalization or Westernization do not necessarily want to turn back time to an age before the internet. Campaigning anti-globalization movements such as Occupy want to curtail the influence of global corporations and the World Bank. Yet they retain a strong belief in global citizenship and use global media networks to spread their anti-capitalist message worldwide.

It may also be true that the case against globalization has been exaggerated. Some academics have suggested that the Milanovic curve has underestimated the financial success of poorer groups in developed countries and that the reasons for the rise of **populism** in Western democracies have far more complex causes than opposition to free trade and immigration alone. Nonetheless, the 2016 summit meetings for the powerful G7, G20 and OECD groups all agreed on one thing: in order to counter tribalization and the rise of disruptive and sometimes dangerous populist movements, there needs to be a greater global effort to reduce inequality and ensure that globalization has benefits for more people and places in the future.

> **PPPPSS CONCEPTS**
>
> Think about how anti-globalization movements may have arisen in different places as a reaction against the same two particular processes: 1) the spread of global culture and 2) the persistence of global inequality.

CASE STUDY

ANTI-GLOBALIZATION PROTEST VOTING IN DEVELOPED WORLD REGIONS

In the USA, opposition to globalization assumed the form of Donald Trump in 2016 – his anti-global rhetoric was one way he convinced sufficient US voters to elect him as their president. In the UK, it takes the form of the UK Independence Party (UKIP) while France's National Front party now has the support of 45 per cent of working-class (or 'blue-collar') voters – but less than 20 per cent of professionals.

These movements share a common aim, which is to 'regain control' of their borders. Theirs is a nationalist philosophy that is relatively lacking in enthusiasm for multiculturalism and internationally minded politics.

Deep schisms in British society were revealed by its referendum on EU membership. The country was not united in its desire to quit. Support for 'Brexit' was high among pensioners, rural communities and urban areas in northern England, whereas younger voters, Londoners and Scots favoured remaining.

In the USA too, the divisions that have emerged are geographically and sociologically complex. Coastal states like California and places with a high proportion of Hispanic voters supported Hillary Clinton's broadly 'business as usual' pro-globalization manifesto in the 2016 election; while some rural interior states and deindustrialized urban areas with the highest proportion of white, poorer and older voters chose Donald Trump to be their leader (Figure 6.13).

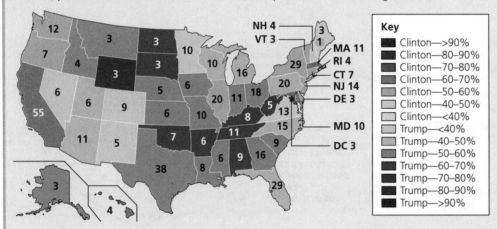

Figure 6.13 The 2016 US presidential election voting map shows the division between more internationally minded US states (Clinton supporters) and those states where a greater proportion of voters would be happy to retreat from globalization (Trump supporters)

CASE STUDY

ANTI-WESTERN MOVEMENTS AND CONFLICT IN DEVELOPING WORLD REGIONS

Nigeria is a deeply divided state. Capital city Lagos is thriving as a highly connected global hub, while a violent campaign against the Westernization of Nigerian society is being led simultaneously by the Boko Haram militia group in the rural northeast. In 2014, 200 schoolgirls were abducted and forced into slavery by Boko Haram, whose name translates loosely as 'Western education is a sin'.

Similar extremist organizations and militias have sprung up in recent years across Asia, the Middle East and Africa – including Iraq, Libya, Syria, Nigeria, Pakistan, Indonesia and Afghanistan. Militias often find support in poorer rural areas where the economic gains of globalization have been few, if any. In their propaganda, these movements often profess to be anti-Western and pro-Islamic (though theirs is very widely viewed as a perverted interpretation of this religion). Many of these local militia groups have forged connections online to create a 'global jihad' movement rooted in a common pseudo-religious identity.

Daesh (alternatively called ISIS or IS) is very effective at using modern communications to brainwash, groom and recruit young Muslims from other states, including EU nations. It is also very good at creating jihadist cells in distant major cities and activating them to kill large numbers of innocent civilians (see Unit 6.3, page 114). European countries also need to work out how to handle the rising number of their own citizens who, having left their own country to fight for various militias and causes, are now battle-hardened and returning home (Figure 6.14).

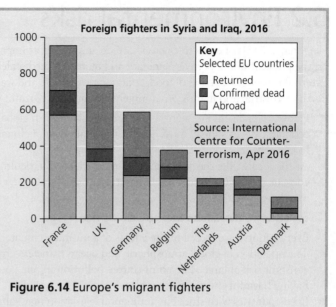

Figure 6.14 Europe's migrant fighters

■ KNOWLEDGE CHECKLIST

- Different kinds of risk for global interactions
- The risks of hacking and identity theft
- How surveillance affects personal freedoms
- The risks businesses are exposed to through their global supply chains
- Political, economic and physical reasons for interrupted global flows
- The impact of the 2011 Japanese tsunami on global supply chains
- The concept of political and economic sovereignty for states
- The economic risk posed by profit repatriation and tax avoidance by TNCs and wealthy individuals
- Disruptive technologies and the risk they pose to state sovereignty, including drones and 3D printing
- The correlation between increased globalization and renewed nationalism and tribalization
- The risk of anti-global tensions in developed countries
- The risk of anti-Western movements and conflict in developing countries

EVALUATION, SYNTHESIS AND SKILLS (ESK) SUMMARY

- How risk exposure increases as societies and businesses become more globalized
- How perspectives vary on whether the advantages of globalization compensate for the risks

EXAM FOCUS

DEVELOP THE DETAIL

Below is a sample part (b) exam-style essay question and a planning table to help you answer the question.

1 Read the question carefully.
2 Reflect on why the planning table is a good way to approach answering this question.
3 Using facts and concepts from this and other units, write one or two paragraphs to support each box contained in the table.
4 Finally, write a summary statement saying whether you believe globalization is still advancing or retreating overall.

Globalization is now in retreat because of the risks it has created. Discuss this statement. (16 marks)

Planning table

Aspects of globalization	Is globalization retreating, pausing or advancing?
Economic globalization	What is happening to TNC supply chains, foreign investment flows and levels of migrant remittances? What role does risk play?
Social globalization	What is happening to international migration flows? What trends can we see in social network growth and internet use? What role does risk play?
Cultural globalization	How are cultures changing locally and globally? How are different cultures reacting to globalization? What role does risk play?
Political globalization	What recent progress has there been for global agreements and international organizations? What are the risks that threaten global political cooperation?

6.2 Environmental risks

Revised

Heightened levels of economic activity, trade and transport impact inevitably on the physical world, both at planetary and more localized scales. The **negative externalities** of global flows of food, commodities, people and waste have affected Earth's terrestrial and marine environments in multiple harmful ways.

- Planetary-scale environmental risks are associated with globalization and global economic development, most notably the challenges of climate change and biodiversity loss.
- At the local scale, the global shifts of industrial and agricultural activities by cost-cutting TNCs have often brought damaging land, water and air pollution to weakly regulated places.

No society is insulated from these escalating risks.

- Every country is exposed to the effects of a warming climate.
- The operation of global **atmospheric and ocean transfers** ensures there is a constant risk of large-scale **point-source pollution** in one poorly regulated place having harmful effects on other places.
- Even countries with strict environmental legislation may suffer occasionally from major pollution events as a result of human error or natural hazards.

> **Keyword definitions**
>
> **Negative externalities** Costs that arise on account of economic activity, including uncompensated-for environmental damage.
>
> **Atmospheric and ocean transfers** The planetary-scale air and water movements that result from the uneven heating of the Earth by the Sun.
>
> **Point-source pollution** The release of pollutants from a single identifiable site such as a factory chimney within one city or country.

Transboundary pollution

Revised

Transboundary pollution has damaging effects for more than one country. It is most likely to occur when:

- polluting activities take place close to a country's border
- atmospheric, ocean or hydrological processes carry pollution in a direction which crosses a state border
- an especially large-scale pollution event occurs.

> **PPPPSS CONCEPTS**
>
> Think about how global flows and interactions bring environmental impacts at contrasting local and global geographic scales.

It is not always the case that all large pollution events are transboundary, however: the Gulf of Mexico oil spill in 2010 only affected the USA. Table 6.7 explores four kinds of transboundary pollution in different regional contexts.

Table 6.7 Examples of transboundary pollution

Forest fires in South Asia	• Every year, toxic haze spreads across Southeast Asia from Indonesian rainforest fires during the dry season. 2015 was an El Niño year, which lengthens and intensifies the dry season. Prevailing winds blew smoke from blazes on the Indonesian island of Sumatra eastwards towards Singapore and Malaysia. Cardiovascular and respiratory disease caused the deaths of an estimated 100,000 people across the entire region, including 6,500 in Malaysia. An additional 500,000 people suffered respiratory illness. • Indonesia's fires are caused by smallholders or plantation owners clearing land for farming or to meet global demand for palm oil production. Singapore has appealed to Indonesia to tackle the problem but regulations are difficult to enforce owing to fire-starting smallholders living in relatively isolated villages scattered across the rainforest region. • Hopefully, the situation will improve in future years: TNCs that buy palm oil, such as Unilever, are under increasing pressure from buyers, investors and environmental groups (including the Indonesian branch of Greenpeace) to ensure their supply chains are environmentally sustainable and do not contribute to transboundary pollution.
Acid rain in North America and Europe	• Acid rain is precipitation with a pH value below the naturally occurring level of 5.6. It can be caused by anthropogenic sulfur emissions, which have risen worldwide from 5 to 180 million metric tonnes since 1860. In heavily industrialized and polluted countries and regions, acidity as high as pH4 may be expected. • Impacts of acid rain for aquatic ecosystems can be highly damaging – especially where local geology lacks any alkaline buffering capacity. Freshwater lakes underlain by silicate granite or quartzite rocks are worst affected, with severe acidification resulting in impaired reproduction and tissue damage for fish, leading to biodiversity reduction. • Taller industrial chimneystacks reduce the intensity of local impacts but instead allow acid rain to have a reduced (but nonetheless significant) effect across a considerably larger area. In the 1950s only two stacks in North America were higher than 180 metres. Now there are hundreds of greater height, the largest being the 380 metre-high copper-nickel Superstack smelter in Ontario, Canada. Taller chimneys have meant that more sulfur dioxide from US power plants in the Ohio Valley is carried by prevailing winds over the border into Canada. The higher the stacks, the further the pollution travels.
Transboundary aquifer pollution in South America	• Spoiling of water quality in transboundary aquifers – of which there are 273 in the world – is another growing environmental concern. • The Guarani Aquifer underlies 1.2 million square kilometres of land shared by Brazil, Argentina, Paraguay and Uruguay. This vital water store is under pressure from many globalized activities such as pulp production and cattle rearing for international markets. Pollution in any one country enters the hydrological system and is transferred to neighbouring states.

CASE STUDY

FUKUSHIMA TRANSBOUNDARY POLLUTION EVENT

The Fukushima Daiichi nuclear plant, 240 kilometres northeast of Tokyo, is one of the 15 largest nuclear power stations in the world. Following the magnitude 8.9 earthquake strike and subsequent tsunami that affected this part of Japan in 2011, a partial nuclear meltdown and several explosions took place at Fukushima. The meltdown was rated at Level 7 on the International Nuclear Event Scale, placing it in the same category as the Chernobyl disaster of 1986, described as a 'major release of radioactive material with widespread health and environmental effects requiring implementation of planned and extended countermeasures'.

The earthquake caused the Fukushima reactors to shut down automatically when their motion sensors felt tremors. However, the cooling system required to remove residual heat from the core failed and the tsunami knocked out the backup generators designed as a last measure to keep cool water pumping. As a result, the uranium heat elements overheated, causing water to evaporate from the system and generating explosive hydrogen, with disastrous effects. This resulted in a pattern of extensive transboundary air, ground and water pollution (Table 6.8). Fukushima radioactive pollution has since spread globally, at extremely low concentrations but traceable using ultra-sensitive instruments.

Table 6.8 Transboundary air, ground and water pollution resulting from the Fukushima disaster

Air and ground pollution	• Radiation escaped immediately into the atmosphere when hydrogen explosions occurred in two reactors and steam vented from the reactor buildings. Increased radiation levels close to the plant reached 400 Millisieverts (mSv) an hour (a chest X-ray involves exposure of 0.02 mSv). A 20 km exclusion zone was imposed, resulting in the forced migration of 70,000 people (to reduce the risk of thyroid cancer). As it fell to ground in Japan, radioactive dust caused local food sources, including milk and spinach, to show radiation levels seven times higher than the legal limit. • Prevailing winds blowing from the southwest carried the greatest radioactive releases northwards and eastwards of the site. The jet stream moved some emissions across the Pacific Ocean towards the USA. Low-level fallout was detected there just five days after the event: monitoring stations along the US west coast detected a spike in concentrations of radioactive iodine, caesium and tellurium, though not at dangerous levels.
Water pollution	• A large amount of radioactive water percolated from the site into groundwater and local coastal waters, threatening fisheries. In the years since the disaster, Pacific Ocean currents have carried caesium isotopes slowly to the west coast of North America. Pollution was detected offshore from British Columbia, Canada in 2013 and off Vancouver Island in 2015. • The highest detected level to date comes from a sample collected west of San Francisco in 2015. However, the level of radioactive caesium isotopes in the sample (11 Becquerels per cubic metre of seawater) was well below levels thought to pose environmental or public health threats.

The response

Around the world, many countries have rethought their nuclear policy as a result of Fukushima:

- Japan itself has begun to overhaul its nuclear power industry and to reappraise the tectonic risk. After the disaster, reactors were left idle but this resulted in a 30 per cent gap in the country's electricity supply that needed to be replaced by fossil fuels. Debate continues in Japan about the role nuclear power should play.
- In Germany, the government responded by accelerating an existing plan to phase out nuclear power, which used to produce almost one quarter of national electricity, by 2020.
- Switzerland has announced it will decommission its five nuclear power plants by 2034. They generate 40 per cent of its energy. The plan is to introduce greater efficiency and more renewables.

The disaster also revived one of the longest-running debates: how safe is it to build nuclear reactors in areas that are seismically active? Around 90 nuclear reactors out of the global figure of approximately 400 are located in areas of significant seismic activity (Figure 6.15). How many of these do you think are potential sites for transboundary pollution?

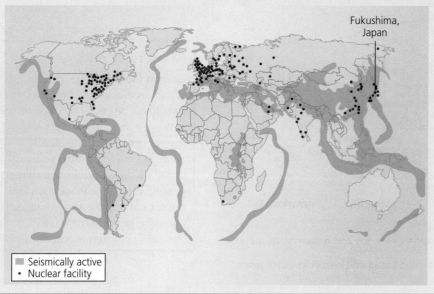

Figure 6.15 Nuclear power stations and seismic activity around the world

Local environmental impacts of global flows

All global flows and movements have some kind of environmental footprint, no matter how small. Even a quick online search using a computer has a tiny **carbon footprint** (an estimated 7 grams of carbon dioxide – around half as much as boiling the water for a cup of coffee). Table 6.9 shows the environmental impacts of activities resulting from several different kinds of global flow.

Table 6.9 The local environmental impacts of different global flows

Global flow	Local environmental impact
FDI and commodity flows	• Some of the world's most polluted sites are global hubs for foreign investment and commodity production. One example is Nigeria's Niger delta. The NGO Amnesty International estimates nearly 7,000 spills occurred during the 1980s and 1990s, harming both people and one of the world's ten most important coastal habitat zones. Large TNCs like Shell, Chevron, Total and ExxonMobil have invested heavily in Nigerian oil, which they export worldwide (Figure 6.16). • Mining operations in Brazil claimed several lives in 2015 when a dam failed at a mine in Samarco owned by BHP Billiton and Vale; the mud and waste slide killed 16 people.
Tourist flows	• Air flight costs have fallen over time, while affluence has risen for many, making travel to distant places more affordable. Expansion of the 'pleasure periphery' – remote regions of the world often possessing wilderness qualities – puts stress on previously undisturbed fragile and unique environments. • Since 2011, for instance, the average daily number of visitors at Machu Picchu has far exceeded the daily limit of 2,500 agreed between Peru and UNESCO. The visitor flow to Machu Picchu has risen from 200,000 to 1.2 million people since 2000; extensive site damage from trampling has now put Machu Picchu on the UNESCO 'endangered' list.
Shipping flows	• Globalization provides invasive species with multiple means of international travel. For instance, the Chinese mitten crab arrived in UK waters as an ecological stowaway in container ships' ballast water. The arrival of invasive species can lead to indigenous species being wiped out. • Island ecosystems are especially vulnerable to indigenous species loss following the arrival of invaders: hitchhiking rats have radically altered food webs in the Galapagos Islands.
Waste flows	• China is the top destination for EU waste flows; in 2010, 7 million tonnes of plastic, 28 million tonnes of waste paper and 6 million tonnes of steel scrap were sent there for recycling. • An estimated 250,000 tonnes of used electrical products flow from the EU to West Africa and Asia each year. Ghana, Nigeria, India, Pakistan and China are major recipients of sometimes dangerous and illegal wastes that, under EU law, ought not to have been exported.

An extensive network of trade flows connects different world regions with the oil-producing Niger delta (one of the world's most heavily polluted sites)– see Figure 6.16.

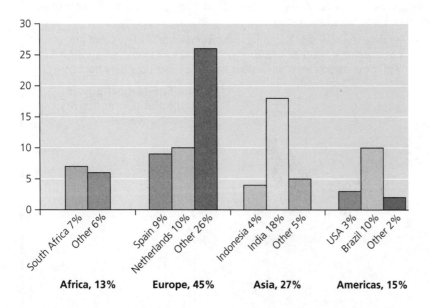

Figure 6.16 Nigeria's crude oil and condesates exports by destination, 2014

Note: Nigerian crude oil and condensate exports averaged 2.05 million barrels per day.

Source: U.S. Energy Information Administration based on Lloyd's List Intelligence (APEX tanker data)

Keyword definition

Carbon footprint The amount of carbon dioxide used by an individual, organization or country as they go about their everyday lives or operations. It is usually measured in terms of the volume of carbon dioxide emitted into the atmosphere as a result of fossil fuel use per unit of time (for example, annually) or per activity (for example, per internet search).

■ Local impacts of marine plastic pollution

Plastic pollution is a problem that has truly 'gone global'. Fragments of plastics washed into the sea by runoff from populated areas have been carried by planetary-scale ocean currents to the remotest corners of the world, including Arctic and Antarctic wilderness areas. The problem has accelerated: more plastic was produced globally in the first decade of the twenty-first century than during the entire twentieth century (the start of which marked the 'birth' of plastic). In 2014, 311 million tonnes of plastics were produced worldwide; this is predicted to rise to over 1,100 million tonnes by 2050. Reasons for the growth in plastic production include:

- Plastic's growing use in everyday life: toothbrushes, credit cards, mobile phones, asthma pumps, Lego bricks, biros, polytunnels and irrigation pipes are all made from plastic.
- The cheap commodity boom driven by low wages in developing and emerging economies has fuelled 'throwaway' attitudes on a global scale: if something breaks, it is often cheaper to 'bin' it than to fix it.
- The boom in bottled water has led to the use of over 2 million plastic bottles every five minutes in the USA. Globally, the figure is far higher. Often, consumption of bottled drinks is driven by 'lifestyle' advertising (given that tap water is perfectly safe to drink in many countries).

Plastic is now believed to constitute 90 per cent of all rubbish floating in the oceans and the UN Environment Programme estimates that every square mile of ocean contains 46,000 pieces of floating plastic. Large areas of the Earth's oceans have become particularly polluted with plastic fragments due to the operation of surface gyres. These gyres are circular currents in the oceans; they move clockwise in the northern hemisphere and anti-clockwise in the southern hemisphere (Figure 6.17). In the North Pacific Ocean, there is now a floating plastic 'garbage patch' that is twice the size of Texas. It is composed of shampoo caps, soap bottles and fragments of plastic bags, in addition to much smaller particles called microbeads which have been needlessly added by TNCs to toothpaste and shower gels (Greenpeace is campaigning to have microbeads banned globally).

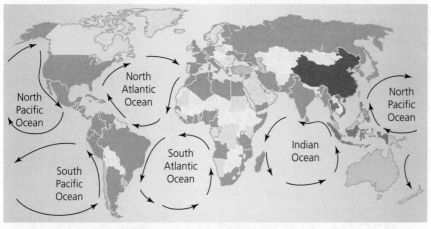

Figure 6.17 Global pattern of source regions for plastic pollution, and gyres that move and trap material

Gyre systems also convey plastic waste to isolated islands and coral atolls far from any pollution source:

- High levels of plastic rubbish have been found on remote Arctic islands over 1,000 kilometres from the nearest town or village, carried there from polluting countries around the world by ocean currents. Muffin Island is one of the most remote places on the planet, yet plastics from Norway, Spain and the USA litter its beaches. This is a pollution problem that does not respect state boundaries.

- Plastic pollution of the Hawaiian Islands, such as Tern Island, has been widely reported by campaigning groups (and provided a stimulus for the recent consumer-led drive to reduce throwaway plastic bag use in the UK).
- Rubber ducks have been washed ashore on once-pristine Alaskan beaches after a floating flotilla of plastic toys was set adrift as a result of a 1992 container ship accident in the Pacific Ocean.

In recent years, scientists have become increasingly concerned with the impacts of plastic pollution on marine species and food webs. Data on seabirds showing the ingestion of plastic waste as being a cause of death first began to appear in the 1950s; 95 per cent of dead fulmars (a common seabird) washed ashore in Scotland will have some plastic debris in their gut. Worldwide, 260 species of bird and mammal are known to ingest or become entangled in plastic wastes. Discarded red lids from water bottles are a particular problem – in size and colour they mimic the appearance of the krill shrimp that albatross eat. Autopsies have shown an abundance of red-coloured debris in the gut of dead albatross. Unit 6.3 (page 108) looks at ways of trying to deal with the challenge of ocean pollution.

Shipping pollution patterns

World trade in oil has been a great economic success story for nations controlling its flow, such as UAE and Saudi Arabia. However, containerized oil movements have brought scores of devastating transboundary pollution events to territories flanking shipping lanes since supertanker technology developed after the 1950s. The coastal margins of both France and the UK were severely affected by 119,000 tonnes of oil released from the *Torrey Canyon* supertanker in 1967, after it struck a reef in the English Channel *en route* from Kuwait to the UK's Milford Haven. This was the first major oil spill to make world headlines:

- Some 15,000 seabirds were killed.
- Around 80 kilometres of UK beaches and 120 kilometres of French coastline were contaminated.
- It remains the UK's worst-ever environmental disaster to date.

Figure 6.18 shows the distribution of the 22 largest spills since then and their proximity to major oil shipping lanes. Some of these incidents, despite their large size, caused little or no environmental damage as the oil was spilt some distance offshore and did not impact coastlines. The same rule that applies to natural hazards is true here also: for an event to become truly hazardous it must affect a populated area.

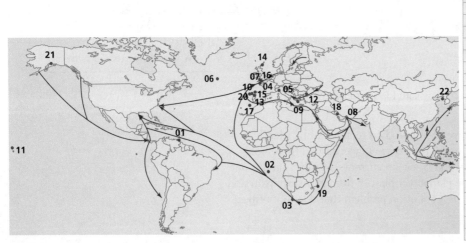

Position	Ship name	Year	Spill size (tonnes)
1	ATLANTIC EMPRESS	1979	287,000
2	ABT SUMMER	1991	260,000
3	CASTILLO DE BELLVER	1983	252,000
4	AMOCO CADIZ	1978	223,000
5	HAVEN	1991	144,000
6	ODYSSEY	1988	132,000
7	TORREY CANYON	1967	119,000
8	SEA STAR	1972	115,000
9	IRENES SERENADE	1980	100,000
10	URQUIOLA	1976	100,000
11	HAWAIIAN PATRIOT	1977	95,000
12	INDEPENDENTA	1979	94,000
13	JAKOB MAERSK	1975	88,000
14	BRAER	1993	85,000
15	AEGEAN SEA	1992	74,000
16	SEA EMPRESS	1996	72,000
17	KHARK 5	1989	70,000
18	NOVA	1985	70,000
19	KATINA P	1992	67,000
20	PRESTIGE	2002	63,000
21	EXXON VALDEZ	1989	37,000
22	HEBEI SPIRIT	2007	11,000

Figure 6.18 The 22 largest oil spills that have occurred since the *Torrey Canyon* disaster in 1967

It is of note that 20 of the largest spills recorded occurred before the year 2000 despite an overall increase in oil trading since the mid-1980s. In the 1990s there were 358 spills of 7 tonnes and over, resulting in more than 1 million tonnes of oil lost; whereas in the 2000s there were 181 spills of 7 tonnes and over, resulting in less than 200,000 tonnes of oil lost. In part this is due to successful **global governance** and the voluntary adoption of the United Nations Convention on the Law of the Sea (UNCLOS) by 156 signatory states.

- One particular achievement has been progressive retirement of the worst offending single-hulled oil tankers that were too easily damaged in the past. The last major single-hulled disaster occurred in 2002 when the *Prestige* supertanker sunk off the Galician coast, leading to the largest environmental disaster in Spain's history.
- UNCLOS legislation makes it illegal for ships that have recently delivered oil to use seawater to wash out their tanks (flushing of tanks has been a significant cause of oil pollution along major shipping lanes).

Yet even when taken out of service, these leviathans perpetuate a lasting legacy of ecological and human harm. Vessels are routinely sent to India, Pakistan and Bangladesh where shipbreaking is carried out at low cost by poorly paid labourers. Toxic materials leak from filthy hulls, polluting coastal waters where the shipbreaking yards are located.

> **Keyword definition**
>
> **Global governance** The term 'governance' suggests broader notions of steering or piloting rather than the direct form of control associated with 'government'. 'Global governance' therefore describes the steering rules, norms, codes and regulations used to regulate human activity at an international level.

■ Carbon footprints for global flows of food, goods and people

Economic activity today takes places on an unprecedented scale. The sheer number of people living their lives as producers or consumers of commodities has brought a step-change in levels of environmental stress. Accelerated cross-border flows of greatly increased volumes of food, goods and people have inflated humanity's planet-wide carbon footprint enormously.

Modern globalization is not entirely responsible for starting this process of carbon footprint growth, of course. It was Western Europe's early Industrial Revolution after around 1750 that really lit the fuse for global atmospheric change (according to prevailing scientific interpretation of Antarctic ice core evidence). However, today's global networks of production and consumption – relying as they do on the perpetual motion of container ships, aeroplanes and mega-trucks filled with mass-produced consumer goods – have clearly accelerated greenhouse gas (GHG) emissions, resulting in the 'hockey stick' scenario that we are all too familiar with (Figure 6.19). It tells of excessive food miles clocked up by agricultural produce and cheap no-frills airlines flying increased numbers of well-to-do pleasure-seekers around the world (in the EU alone, carbon emissions from air flight have tripled since 1990).

> **PPPPSS CONCEPTS**
>
> Think about the spatial interactions between developed and developing countries. Should the place and community that consume goods take some responsibility for the carbon emissions of the place that produces them?

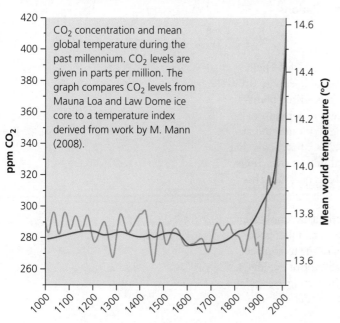

Figure 6.19 'Hockey stick' trends for CO_2 and average world temperature

The carbon footprint concept is a useful way to start thinking about how the activities of one individual or society may potentially affect others. However, the validity and reliability of carbon footprint measures are worth reflecting on critically, as Figure 6.20 suggests.

Increasing scale		
	Personal carbon footprint	• An individual's footprint can be calculated over the course of the year by estimating the embedded carbon in: 1) manufactured goods purchase (e.g. smartphones and TVs; 2) air flights (a return flight from London to Malaga emits 400 kg of CO_2 per passenger); 3) home heating (one extra degree on a household thermostat accounts for an additional 25 kg of CO_2 per person per year); 4) annual per capita consumption of food and drink, including carbon emissions linked with food production, transport and cooking; 5) commuting (e.g. an individual's share of the carbon footprint of each train journey taken). • *Many other factors and activities could be included: but can we agree on what they are? How accurately can someone's carbon footprint be measured?*
	Corporate carbon footprint	• A business can attempt to estimate the carbon footprint of all of its business premises but should it include employees' journeys to work, or is that the responsibility of the worker? Should carbon emitted by a TNC's overseas operations be included, or just emissions in the country of origin? Are the carbon emissions of outsourced goods the responsibility of the TNC that orders and buys them? Should the carbon embedded in every single component used in the manufacturing of a car or television be included in the final footprint estimate for the completed product? • *In 2010, UK supermarket Tesco abandoned an attempt to introduce carbon footprint labelling on ready meals and other goods because the number of ingredients and factors to be considered made the task too complex.*
	National carbon footprint	• Emerging economies have fast-growing emissions. China now contributes the greatest share of world emissions (25 per cent) while accounting for 20 per cent of the world population. Its emissions are still rising by 8–9 per cent annually, primarily due to higher coal consumption. However, improvements in energy efficiency and greater use of renewables has meant that the rise has not been as high as it could have been. China's carbon intensity – the amount of CO_2 emitted per unit of GDP – is falling. • Many of the world's least developed countries, such as DRC, continue to make a negligible contribution to anthropogenic GHG emissions. • Developed countries have high but falling emissions according to the data their governments produce. However, these figures only reflect 'domestic' emissions (gases emitted within each state's borders) and exclude 1) the operations of their own TNCs' overseas operations and 2) any of the emissions of the countries they import from. As a result, the UK government, for example, has claimed responsibility for 1.5 per cent total global emissions annually. • *Critics say that this is a disingenuously low estimate because UK citizens are major consumers of carbon-emitting goods and services produced elsewhere in the world.*

Figure 6.20 Thinking critically about carbon footprint calculations

Despite still being high emitters when compared to the world's poorest countries, many of the world's developed countries are now reducing their carbon footprint size in part owing to a move towards renewable energy (in the case of Germany) or a shift from coal to gas burning (a gas-fired plant produces half the emissions of a coal-fired one). For this and other reasons, it is increasingly common to see the world's developed countries situated at position 4 on the environmental Kuznets curve (Figure 6.21). This shows that these countries have begun to reduce their environmental impact over time. However, the study of global interactions leads us logically to question this hypothesis:

- Falling domestic carbon emissions in some developed countries mask the fact that these states now import much of their food and consumer goods from other countries since deindustrialization and global shift took place.
- Developed countries are reducing their domestic carbon footprint but are also, in reality, increasing their carbon consumption by importing ever-greater volumes of energy-intensive food and goods from other countries.

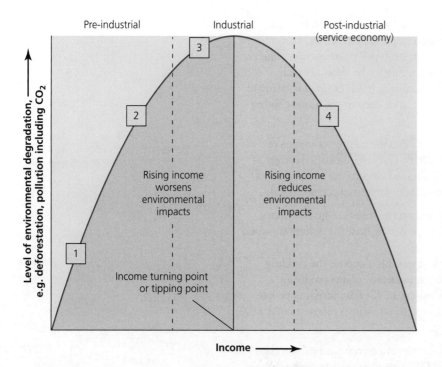

Figure 6.21 The environmental Kuznets curve – is this a valid indicator of the carbon footprint size of individuals and societies?

Key
1. Europe pre-Industrial Revolution, remote Amazonia today, Bangladesh pre-1970s
2. Bangladesh today, China in the twentieth century
3. China today
4. Developed countries today

Global shift and the environment

Transnational Corporations (TNCs) seek low-cost sites for their manufacturing and refining operations wherever possible. Cheap labour is often a key locational consideration, underlying **global shift**, as we learned in Unit 5.1. However, another attractive location factor can be weak environmental controls. In high-income nations, bodies such as the UK Environment Agency are well funded to carry out their brief of closely monitoring industrial operations. In comparison, far less 'red tape' exists in many developing countries and emerging economies.

Keyword definition

Global shift The international relocation of different types of industrial activity, especially manufacturing industries. Since the 1960s, many industries have all but vanished from Europe and North America. Instead, they thrive in Asia, South America and, increasingly, Africa.

■ Polluting manufacturing industries

Although it happened more than 30 years ago, the pollution event that occurred in the Indian town of Bhopal remains highly relevant to the contemporary study of global interactions. There are two reasons for this:

- Justice has not yet been won for the people who were affected. The people of Bhopal have yet to receive adequate compensation for their exposure to one of the worst industrial disasters on record. In 1984, the US company Union Carbide's poorly maintained Indian pesticide plant accidentally released a lethal plume of toxic gas, killing thousands. Four key safety measures failed that night, including the plant's cooling system and its flare tower, which should have burned off the gas before it fell to earth. At least 3,000 people died on the night of the gas leak, while hundreds of thousands more fled Bhopal – a *forced migration*. Many of these migrants may have died shortly afterwards, leading Amnesty International to suggest that the true immediate death toll on the night was nearer 7,000.
- The slow progress made towards achieving compensation is due in part to the later acquisition of Union Carbide by Dow Chemical. Acquisitions are a routine way for successful TNCs to build market share, power and influence, as Unit 4.2 explains. They also provide the perfect pretext for legal evasion: for many years, Dow Chemical claimed no direct responsibility for Union Carbide's past actions in India. The company has resisted making large pay-outs to the Bhopal petitioners, and has also failed to clean up the site. In 2016, survivors gathered in Bhopal and burned an effigy of US President Barack Obama to protest against his government continuing to 'shield' the TNC Dow Chemical.

In more recent years, the breakneck speed of manufacturing growth in China has given rise to a pollution problem on a much larger scale that has potentially affected hundreds of millions of people. The 'airpocalypse' phenomenon reduces Chinese life expectancy by up to five years according to the World Health Organization (WHO). The culprit is very high average levels of small particulate matter known as PM2.5. These deadly particles settle deep in the lungs, causing cancer and strokes.

Increasingly, civil society organizations inside China are raising awareness of the issue by staging 'not in my back yard' (NIMBY) protests organized using social media networks (as Unit 3 explained, Chinese citizens are active users of 'shrinking world' technology despite a relative lack of external connectivity).

- In 2010, more than 1,000 villagers marched on the streets of Jingxi county, Guangxi Zhuang Autonomous Region, to protest against the pollution caused by an aluminium plant.
- Many of the 'NIMBY' protests taking place in China oppose the building of new petrochemical and plastics factories, especially plants producing paraxylene, or PX (used in paints and plastics). In 2013, hundreds of people gathered in Kunming, Yunnan province, to protest against plans to build a PX plant in a nearby town.

The Chinese government pledged recently to improve environmental standards and to tackle industrial pollution. The 13th Five-Year Plan (FYP) covers China's economic development over the period 2016–21 and includes legislation for a 25 per cent reduction in factory emissions of PM2.5.

Figure 6.22 Air pollution in China

■ Agribusiness systems

Though not always generating the same kind of dramatic headlines as chemical spills or toxic gas, the ecological transformation of 40 per cent of Earth's terrestrial surface into productive agricultural land – much of it now in the hands of major **global agribusinesses** – is a profound environmental change. Food chains and nutrient cycles have been modified across the globe, often bringing habitat loss and severe biodiversity decline on a continental scale.

The larger agribusinesses, such as Del Monte, are TNCs whose extensive production networks are responsible for the truly diverse range of food source regions that a typical European, North American or East Asian shopper encounters in the supermarket aisles. When measured in terms of their controlling land interests, the largest firms exhibit enormous size and power (Table 6.10). Their impacts penetrate deeply into some of the poorest societies of the world, such as East Africa and southern Asia (Figure 6.23). A variety of activities, including the intensive production of cash crops, cattle-ranching and aquaculture, bring many damaging environmental effects (Figure 6.24).

> **Keyword definition**
>
> **Global agribusiness**
> A transnational farming and/or food production company. This blanket term covers various types of TNC specializing in food, seed and fertilizer production, as well as farm machinery, agrochemical production and food distribution.

Table 6.10 Some large agribusinesses and their activities

Name	Headquarters	Activities
Kraft Foods	Illinois, USA	Food processing
Unilever	Rotterdam, Netherlands	Conglomerate
Nestlé	Vevey, Switzerland	Food processing
Carrefour	Levalloir-Perret, France	Food retail
Monsanto	Missouri, USA	Agricultural technology
Cargill	Minnesota, USA	Food production and technology

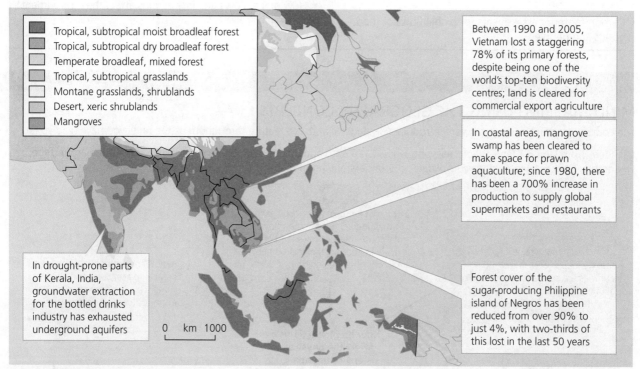

Figure 6.23 Impacts of agriculture, food and drink TNCs operating in parts of Asia

Eutrophication
Around 20 major 'marine dead zones' lie scattered around the world's coastal margins. These are sites where intense inputs of fertilizers in agricultural runoff over-stimulate ecosystem productivity. This results in algal bloom growth and its subsequent collapse, leading to de-oxygenation of water and species death. Some of the worst-affected areas are hub regions for global agribusiness, such as the Gulf of Mexico.

Biodiversity loss
With so much of the world's land surface used for farming, many of Earth's 1.4 million identified species have experienced habitat loss. Increasing numbers are on the endangered list. The largest agribusinesses have often promoted wheat, maize, rice or potato monoculture (both shaping and reflecting the homogenization of diet that is an aspect of cultural globalization). These four crops now account for 60 per cent of plant-derived calories in human diets worldwide. Simultaneously, just ten key animal species now provide 90 per cent of all meat eaten. Global genetic diversity is vital for lasting ecological sustainability but is clearly threatened in many ways. Where monoculture is practised, cash crop farming can also lead to declining soil fertility and increased soil erosion.

Forest services loss
Forests deliver vital ecosystem services for people and places. For instance, tropical rainforest provides interception cover that naturally limits storm runoff. Rising incidence of flooding in the Ganges delta can be linked to timber removal, with much of the wood feeding global hardwood demand. Mangrove forest similarly offers protection to tropical coastal margins – in this case, against storm surges. However, insatiable world consumer demand for tiger prawns directly leads to mangrove clearance in places like Indonesia and Madagascar. Space is created for prawn aquaculture ponds but at the cost of natural tsunami protection.

Water scarcity
The worst effects occur when intensive crop farming is introduced to areas where only limited water supplies are naturally available. Thirty farms growing flowers destined for European supermarkets have contributed to the shrinkage of Kenya's Lake Naivasha. Groundwater abstraction by Coca-Cola may have contributed to water scarcity in parts of Indian Kerala. The term 'virtual water' is used to describe the water that has been 'embedded' in the production of food or goods for global markets. Each EU citizen consumes around 4,000 litres of virtual water every day, according to one estimate, on account of their lifestyle.

Figure 6.24 Selected environmental impacts of agribusiness activity

■ KNOWLEDGE CHECKLIST

- The concept of transboundary pollution with supporting examples
- One case study of transboundary pollution (TBP), including the consequences and possible responses (Fukushima)
- The localized environmental impacts of different global flows
- Localized pollution impacts along shipping lanes
- The carbon footprint concept
- Carbon footprint assessments for global flows of food, goods and people
- Environmental issues linked with the global shift of economic activity
- The global shift of polluting manufacturing industries
- The environmental impacts of global agribusinesses and their food production systems

EVALUATION, SYNTHESIS AND SKILLS (ESK) SUMMARY

- How different global interactions affect physical environments and processes
- How global flows and interactions affect the environment at varying scales

EXAM FOCUS

MIND-MAPPING USING THE GEOGRAPHY CONCEPTS

The course Geography Concepts were introduced on page vii and also feature throughout all the units you have read so far.

Use ideas from Unit 6.1 (and, where relevant, Units 4.1–5.3) to add extra detail to the mind map below (based around the Geography Concepts) in order to consolidate your understanding of global environmental risks.

How do global flows affect the physical environment?

- **Interactions**: Different kinds of interaction vary in their impact; information flows may cause the least harm and may even help improve the environment (e.g. online campaigning by CSOs)
- **Place**: Some local places in developing countries are severely affected by the polluting flows associated with global shift; other places are cleaner as a result of global shift
- **Scale**: The physical environment is affected at the local (place) scale and also the global scale (on account of the carbon footprint of different flows)
- **Power**: Impacts vary greatly from place to place due to the power of governments and other stakeholders to shape effective legislation to prevent pollution; some governments may feel powerless to pass strict laws because they do not want to deter flows of foreign investment
- **Possibility**: Climate change predictions and possibilities vary enormously due to the complexity of global interactions and uncertainty over future economic growth and globalization trends
- **Processes**: Atmospheric, ocean and river processes all complicate the pattern of pollution impacts; point-source emissions become more widely distributed for example

Planning an essay

Below is a sample part (b) exam-style question. Use some of the information from your mind map, or your own ideas, to produce a plan for this question. Aim for five or six paragraphs of content; each should be themed around a different key point you want to make, or a particular concept or case study. Once you have planned the essay, write the introduction and conclusion for the essay. The introduction should define any key terms and ideally list the points that will be discussed in the essay. The conclusion could evaluate the relative importance of the key factors and justify why any factors are especially important for particular geographic contexts, scales or perspectives. The levels-based mark scheme is on page vii.

Examine how different kinds of global interaction can affect the physical environment. (16 marks)

6.3 Local and global resilience

Revised ☐

Faced with so many potential risks to 'business as usual' globalization, what, if anything, can different stakeholders do to protect themselves? Figure 6.25 shows the broad choices that are available. The first of these – managed retreat – is mirrored in the populist movements that have sprung up around the world (Unit 6.1). But would people who wish for greater barriers to migration and trade be prepared to give up the internet and access to cheap imported food and goods? To be clear: a complete retreat from globalization would bring many costs.

The other options – adaptation and mitigation – involve accepting that globalization is here to stay. However, efforts can be taken to protect ourselves from its negative externalities. Mitigation efforts to reduce international migration could involve greater efforts to end poverty and conflict in those places which generate the largest volumes of migrants. We can adapt to the global threat of computer viruses by installing anti-virus software.

An important part of adaptation and mitigation work involves:
- becoming aware of where new risks have developed as result of the way human activity has spread globally and in complicated and networked ways
- modifying global systems in ways that reduce these new risks or provide alternative networks.

Together, these strategies are building **resilience**. Resilient systems have an ability to 'bounce back' if a shock does occur, such as the Global Financial Crisis or the 2011 Japanese tsunami.

> **Keyword definition**
> **Resilience** The capacity of individuals, societies, organizations or environments to recover and resume 'business as usual' functions and operations following a hazard event or other system shock.

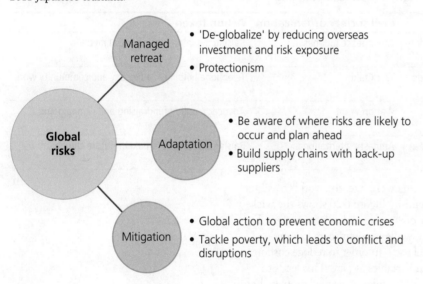

Figure 6.25 Coping with global risks

Managing global risks and wicked problems

Revised ☐

The management of risk to society is, in large part, the responsibility of the state. This is certainly true of natural hazard risks: the USA's Federal Emergency Management Agency (FEMA) is funded with billions of dollars by the state to help protect people from the impacts and after-effects of hurricanes and earthquakes.

In the views of many people, however, governments have often been slow to recognize and mitigate many of the risks to global systems which Units 6.1 and 6.2 explored. This is because of laissez-faire attitudes towards economic globalization favoured by neoliberal governments and financial institutions. Freedoms have been engineered for TNCs to invest worldwide and build extensive supply chains, using free trade zones and tariff-free MGOs. Companies have been allowed to build the necessary global architectures to construct a shrinking world using fibre optic cables and satellites. Yet the problem with leaving so much of this work to market forces is that the new risks and problems that multiply in these systems are not always recognized by governments until it is far too late: the GFC is the ultimate example of this.

International civil society organizations (CSOs) can have a vital role to play here as a result. Citizen-led campaigning groups have often played a critical role in:

- uncovering new environmental and social risks associated with global interactions
- raising awareness about, and proposing solutions for, these risks
- taking action to pressure powerful state governments, MGOs and TNCs into acting to mitigate these risks through the adoption of new rules, agreements, frameworks and legislation.

Table 6.11 shows examples of the work of civil society organizations (CSOs) in relation to some of the global risks and issues you have read about in this book. Not all of these campaigns have achieved their goal yet, however. This is owing to the sheer complexity of some risks and issues, which result in them becoming **wicked problems**.

> Keyword definition
>
> **Wicked problem** A challenge that cannot be dealt with easily owing to its scale and/or complexity. Wicked problems arise from the interactions of many different places, people, issues, ideas and perspectives within complex and interconnected systems.

Table 6.11 Examples of civil society organization (CSO) campaigning

Issue	Civil society organization	Action taken
Indonesian forest fires and transboundary pollution (Unit 6.2)	Greenpeace Indonesia	Raising awareness using social media
Child soldiers in Democratic Republic of the Congo (Unit 5.3)	War Child	Telephone helplines, advocacy and community work
Exploitation of Bangladeshi textile workers (Unit 5.1)	War on Want	Advocacy, online fundraising and campaigning
Damage done to Ogoni people's land by oil spills in Nigeria (Unit 5.1)	Amnesty International	Raising awareness (#makethefuture campaign)

At a global scale, wicked problems include climate change, the world's fossil fuel dependence and new forms of political extremism. Figure 6.26 shows the wicked problem of fossil fuel use, which civil society organizations have campaigned against (Figure 6.27). As you can see though, while it might seem 'common sense' for more governments to ban fossil fuel use – in order to reduce carbon emissions – there are a huge number of other variables at play. This leads to inertia by governments as there is no clear path to take. A wicked problem has, by definition, no clearly visible solution.

Wicked problems arise in more localized contexts too: attempts to introduce simple changes – such as improved safety in Bangladesh factories – require citizens, governments, TNCs and outsourcing companies to act; but if safer factories bring higher costs, TNCs may need to take their business elsewhere (you may remember there were similar unintended consequences arising from the introduction of the Dodd–Frank law in Unit 6.1). This is another hallmark of a wicked problem: complex interdependency among the different elements of the challenge may even mean that the 'solution' actually exacerbates the original problem or creates new ones.

The case studies included here highlight cases where campaigning action by CSOs has been at least partly successful in engaging with environmental and social risks that have become wicked problems.

6.3 Local and global resilience

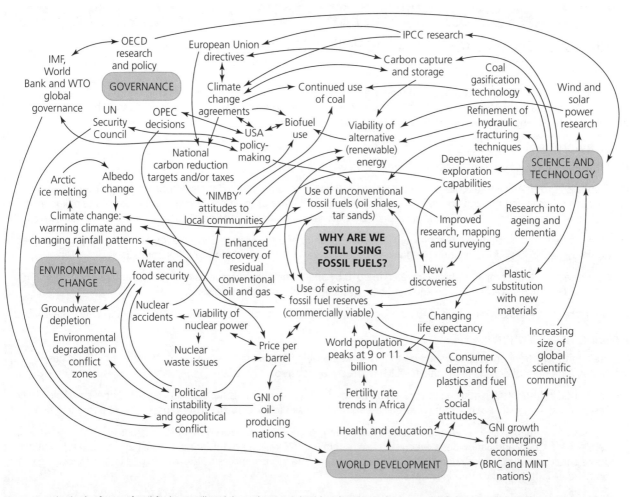

At some point in the future, fossil fuel use will end, but when? Might it be already too late to avoid dangerous climate change? How can we abandon oil sooner without harming the interests of powerful energy companies and oil-producing nations? A complex nexus of interrelated political, economic, physical and technological factors underpins this wicked problem.

Figure 6.26 The wicked problem of continued fossil fuel use

Figure 6.27 An anti-fossil fuels campaign

CASE STUDY

TAKING ACTION TO TACKLE PLASTIC POLLUTION IN THE OCEANS

The issue
The environmental hazards created by worldwide use of throwaway plastic are enormous, as Unit 6.2 (page 97) explains. Plastic pollution is an archetypal wicked problem insofar as it appears to be an insurmountable challenge: the issue seems too big in scale for any single action or organization to make a difference.

How awareness was raised
Nevertheless, various CSOs have attempted to raise awareness of the issue:

- Campaign group Adventure Ecology built a boat called 'Plastiki' made from 12,500 plastic bottles. They sailed it across the Pacific Ocean and through the garbage patch. This caught the eye of the media, raising awareness of the pollution problem. During their voyage in 2010, the expedition crew released videos of the plastic garbage patch onto the internet.
- Many more CSOs – including Greenpeace, The Ocean Cleanup and the Marine Conservation Society – have campaigned on the issue; important foci include the dangers posed by plastic bags, bottle tops and microbeads.
- *Plastic Bag* is a short propaganda film created in 2010 by an international team including American director Ramin Bahrani, Germany's Werner Herzog and members of the Icelandic rock band Sigur Rós.
- There are numerous CSOs dedicated to banning the sale of bottled water in countries where clean tap water is available.

Arriving at a solution
We are a long way from arriving at a solution to this wicked problem. However, the raising of awareness by CSOs has resulted in a number of actions being taken which represent a first step.

- The plastic industry itself is starting to take action by developing new materials such as biodegradable or even edible plastic. Research has shown that a milk protein called casein could be used to develop an edible, biodegradable packaging film.
- Various governments have taken action to ban plastic bags or microbeads. Government restrictions on throwaway plastic bags exist in China and Bangladesh, where the use of thin (<0.025 millimetre thickness) plastic bags has been prohibited (these small bags also block watercourses and sewers during the monsoon season). By law, plastic bags cannot be given away freely in the UK any more in larger shops.
- The USA has banned microbeads from 2017; many global retailers are already removing them voluntarily from their own products.
- The Ocean Cleanup CSO has raised money from its global network of supporters – using an online **crowdfunding** platform – to build a €1.5 million prototype floating barrier made of rubber and polyester, which can catch and concentrate debris. Nicknamed 'Boomy McBoomface', it was launched off the coast near The Hague in 2016 (Figure 6.28). The aim is to upscale the model to produce 100-kilometre V-shaped barriers positioned in the Pacific gyre.
- While no country has considered seriously a ban on bottled water, growing numbers of people worldwide now eschew its use.

Evaluating the action
The size and scale of this problem makes it an enormous global challenge. However, it also means that there are an enormous number of global stakeholders who want to fix the problem. Several key unanswered questions make this an enduring wicked problem, however:

- Will the plastics industry reform itself voluntarily or is global regulation needed?
- Even if new flows of plastic are reduced, what can be done about the enormous existing stores of plastic that have collected already in the gyres? Even if the 'Boomy McBoomface' solution works, what will be done with all of the plastic once it is collected? Can all this be done without harming marine wildlife?
- Plastic use is projected to quadruple by 2050: will any actions we take be 'too little, too late'?

Figure 6.28 A prototype floating barrier designed to capture plastic waste

---

Keyword definition

Crowdfunding Raising sums of money for a good cause or innovation by asking a large number of people to donate a small amount each using an online platform.

CASE STUDY

TAKING ACTION TO HELP FEMALE FARM WORKERS IN SOUTH AFRICA

The issue

In 2005, Gertruida Baartman was working as a fruit picker at a South African farm near Cape Town that supplies European supermarkets. Until recently, she was paid South Africa's 'minimum' wage – which is still less than a 'living' wage – around US $130 per month. The single mother told a newspaper: 'My four children do go hungry but I try my best. I have to pay school fees and sometimes that's a struggle because the fees are high. The school uniforms are expensive for me too and I don't have money to buy them shoes.' Two out of three insecure seasonal workers in South Africa are black women, often lacking the same benefits as men, who are more likely to be on permanent contracts.

How awareness was raised

UK-based CSO ActionAid heard of the plight of South African fruit pickers, thanks to its connections with Sikhula Sonke, a women-led trade union of farm workers in South Africa. ActionAid flew Gertruida to the 2007 annual shareholder meeting of the UK-based TNC Tesco in London. She received a standing ovation from the shareholders, who were horrified to hear about conditions at the base of their value chain.

Arriving at a solution

After Tesco representatives visited Gertruida's farm, there have been improvements, such as a toilet in the orchard where she works and a reduced pay gap between men and women. In 2012, Sikhula Sonke went on to win a 50 per cent increase in the minimum wage for casual farm labourers in South Africa as part of prolonged strike action.

Evaluating the action

This case study is useful because it focuses on an ethical issue which was tackled effectively by a range of stakeholders in different places interacting with one another to shape a positive outcome (Figure 6.29). The power to bring about change was distributed throughout a network of different individuals and organizations.

- Gertruida lacked financial power and was, potentially, another voiceless labourer. Yet she clearly had the drive and determination needed to work towards achieving a better outcome for her family when the opportunity was offered.
- The CSO ActionAid has limited financial resources, which need to be used sparingly and wisely in order to help bring about change. In this example, it facilitated a meeting between the two polar extremities of the Tesco value chain: labourers and shareholders. Consequently, Gertruida and the shareholders met and gained an understanding of one another.
- Finally, the shareholders possess the financial and regulatory power needed to make change (in relation to their own supply chain). Yet they lacked knowledge of working conditions among their own subcontractors. Once they learned of the injustice (and risk to their own company's reputation), they took action.

The final outcome is a positive one, though clearly there is a long way to go: pay for South African farm workers is still terribly low.

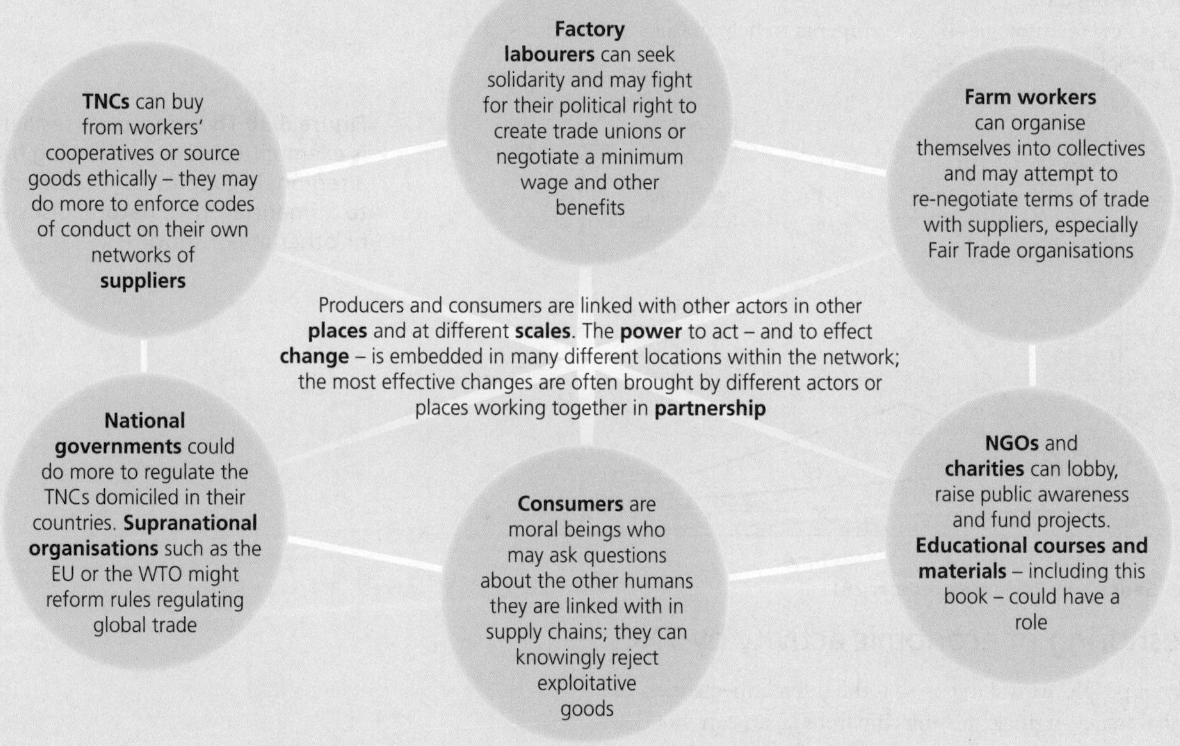

Factory labourers can seek solidarity and may fight for their political right to create trade unions or negotiate a minimum wage and other benefits

TNCs can buy from workers' cooperatives or source goods ethically – they may do more to enforce codes of conduct on their own networks of **suppliers**

Farm workers can organise themselves into collectives and may attempt to re-negotiate terms of trade with suppliers, especially Fair Trade organisations

Producers and consumers are linked with other actors in other **places** and at different **scales**. The **power** to act – and to effect **change** – is embedded in many different locations within the network; the most effective changes are often brought by different actors or places working together in **partnership**

National governments could do more to regulate the TNCs domiciled in their countries. **Supranational organisations** such as the EU or the WTO might reform rules regulating global trade

Consumers are moral beings who may ask questions about the other humans they are linked with in supply chains; they can knowingly reject exploitative goods

NGOs and **charities** can lobby, raise public awareness and fund projects. **Educational courses and materials** – including this book – could have a role

Figure 6.29 Different stakeholders can work together to bring about change

> **PPPPSS CONCEPTS**
> Think about the development gap within South Africa. Who has the power and responsibility to improve pay and working conditions for people like Gertruida?

Strategies to build resilience

When the Philippines were struck by Typhoon Haiyan in 2013, global news reporters praised the 'resilience' of local people. Two years earlier, dire predictions were issued that global supply chains might 'lack resilience' after Japan's devastating tsunami wiped out factories supplying vital parts to American and European manufacturers. In the wake of continuing global financial shocks and acts of terrorism, governments around the world now assert regularly that their citizens must unite to become 'a resilient nation'.

The concept of 'resilience' features in all these examples: it means having the capacity to leap back or rebound, following a disruption or disaster (Figure 6.30). The word's roots lie principally in ecology (analysing the self-restorative power of damaged ecosystems). Academics, business leaders and politicians now embrace the word as a catch-all way of characterizing the capacity of societies, economies and environments to cope with the diverse risks brought by global interactions and human development.

Approaching resilience from a governance perspective, capacity-building is increasingly seen as a central plank of risk management by global and national government agencies and businesses. Some threats cannot be fully mitigated, for example Pacific tsunamis, or (so far) economic 'boom and bust' cycles. In such cases, residual risk remains. Government and business leaders – acknowledging that they cannot guarantee absolute protection – increasingly recognize that they must also do their utmost to ensure that the systems and people they manage can withstand disruption, absorb disturbance, act effectively in a crisis and adapt to changing conditions.

Examples of resilient behaviour that can minimize risk, damage or recovery time include:

- the **reshoring** of economic activity by TNCs
- use of crowdsourcing technologies by government and civil society to generate valuable **big data**
- use of new technologies by governments to help manage global flows of data and people.

Keyword definitions

Reshoring Also known as onshoring and backshoring, this involves a TNC abandoning lengthy supply chains and instead returning productive operations to the country where it is headquartered. The company will no longer make use of a spatial division of labour.

Big data Large data sets that may be analysed using computers. Analysis of big data may reveal new patterns, trends, associations or risks that do not show as clearly in smaller-scale information and studies.

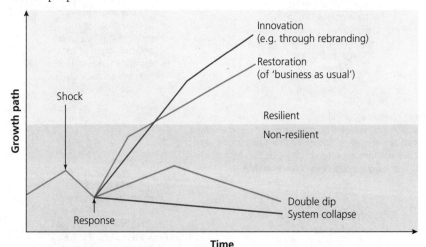

Source: Geography Review November 2011

Figure 6.30 The concept of resilience is exemplified here by showing how a regional economy could respond to a financial crisis, natural disaster or other major shock

■ Reshoring of economic activity by TNCs

> 'What people are waking up to is the interconnectedness of global trade – a single missing chip from Japan can shut down an (American) Ford factory on the other side of the world.'
>
> Richard Ward, Chief Executive of Lloyd's of London